Telegraph and Data Transmission over Shortwave Radio Links

Fundamental Principles and Networks

By Lothar Wiesner

Third edition

SIEMENS AKTIENGESELLSCHAFT
JOHN WILEY & SONS

British Library Cataloguing in Publication Data

Wiesner, Lothar:
Telegraph and data transmission over shortwave radio links.—3rd ed.
1. Radio, Shortwave
I. Title II. Fernschreib-und Datenübertragung über Kurzwelle. *English*
621.3841'51   TK6564.S5

ISBN 0-471-90599-2

CIP—Kurztitelaufnahme der Deutschen Bibliothek

Wiesner, Lothar:
Telegraph and data transmission over shortwave radio links: fundamental principles and
networks/by Lothar Wiesner.–3 ed.–
Berlin; München: Siemens-Aktiengesellschaft, [Abt. Verl.]; Chichester: Wiley 1984—

Dt. Ausg. u.d.T.: Wiesner, Lothar: Fernschreib-und Datenübertragung über Kurzwelle
ISBN 3-8009-1405-0 (Siemens AG)
ISBN 0-471-90599-2

Title of German original edition:
Fernschreib-und Datenübertragung über Kurzwelle: Grundlagen und Netze
Von Lothar Wiesner
4. Auflage–Berlin, München: Siemens Aktiengesellschaft 1984
ISBN 3-8009-1391-7

ISBN 3-8009-1405-0 Siemens AG, Berlin and München

Preface to third edition

In the middle of the last century the electric printing telegraph was invented, which permitted written information to be transmitted from one place to another. Out of this printing telegraph grew the teleprinter in its present form.

In 1888, Heinrich Hertz proved with his experiments the existence of electromagnetic waves that are capable of propagation over great distances. The problem of wireless communication was tackled by Marconi, among others, who succeeded in 1897 in transmitting information over a radio circuit. Not more than 60 years ago, long-range communication was only possible on open-wire lines and cables. This makes us realize the tremendous progress in the field of radio telegraph engineering within the last decades.

In the early days of telegraphy the Morse key and recorder, and later on the teleprinter, were used on physical circuits. Analogously, digital message transmission over radio circuits began with manual Morse operation, followed by high-speed Morse operation employing amplitude modulation and finally by the teleprinter service employing frequency modulation. The Morse key was used for keying the radio frequency of a transmitter on and off in the cadence of the telegraph signals. At the receiving end the incoming signals, rendered audible by a radio receiver, had to be reconverted into the original symbols. For the transmission of binary signals emitted by a teleprinter or similar data source, it was necessary to find new solutions to the technological problems, since interference along the transmission path tends to mutilate the message signals. Great distances can be spanned with radio transmitters having an output of up to 100 kW so that any country in the world can now be reached over intercontinental radio links. Another important step forward was the multiple utilization of radio links with the aid of voice-frequency telegraph terminals.

In the last 25 years the transition from amplitude modulation to frequency modulation, the introduction of diversity reception, and the protection of messages in accordance with the error-correction principle proposed by van Duuren (ARQ system) brought about continual improvement in transmission quality. In the case of frequency modulation the high-frequency stability of the radio equipment now permits operation at a bandwidth matched to the telegraph speed. Intercontinental telex service imposes increasing demands on the transmission quality of a radio link in order to ensure rapid and reliable establishment of connections between the subscribers. Today, telex connections over shortwave links without ARQ protection would be quite out of the question.

5

In spite of submarine cable connections and the present satellite links between continents, the shortwave holds its place as the bearer of continental and intercontinental communication circuits, e.g. for embassies, meteorological services, news agencies and air traffic control authorities. A number of domestic shortwave services may also be mentioned in this connection. They have been set up chiefly by security authorities, by the military and by private companies. It is in this field in particular that a marked increase has been achieved within the last few years in telegraph and data traffic, mostly over small radio stations which offer a maximum of economy.

To ensure smooth interoperation with other countries and continents, a number of technical and operational recommendations have been issued upon international agreement. These recommendations have been worked out by international advisory bodies which have been integrated in the CCITT (Comité Consultatif International Télégraphique et Téléphonique) and in the CCIR (Comité Consultatif International des Radio-communications). International co-operation in the field of radio transmission is promoted by the UIT (Union Internationale des Télécommunications).

This book is intended to familiarize a wide audience with telegraph and data communication over shortwave links. It describes the technical facilities now available for the attainment of maximum reliability in message transmission. To begin with, the physical phenomena encountered in shortwave propagation and the fundamental problems of telegraph communication over radio links are discussed. The characteristics of short waves, types of modulation, and transmission methods as well as diversity and data protection methods are then dealt with in great detail.

In revising the general chapters, emphasis has been placed on increasing the clarity of certain passages. The additions cover in particular the fields of ARQ and FEC systems and of data terminals.

The author wishes to express his thanks to all those whose valuable contributions and suggestions have helped him in writing this book or who have assisted him with the editorial work.

Munich, September 1984

Siemens Aktiengesellschaft

Contents

8

1 Communication Over Radio Links

1.1 Wave Bands

Message transmission by means of radio waves depends on the physical phenomenon that electrical oscillations can be radiated into space and transported over great distances in the form of electromagnetic waves (Fig. 1).

The HF oscillations generated in a radio transmitter are modulated with the message to be transmitted, applied to an antenna and radiated by the latter.

In the field of message transmission the wavelengths of the radiated oscillations are subdivided, according to their different properties, into a number of wave bands or frequency ranges (Table 1).

Messages may be radiated within a band ranging from about 10 kHz to frequencies far beyond 3000 MHz. Within this overall range each subrange is characterized by its peculiar propagation conditions which must be taken into account in the planning of communication systems.

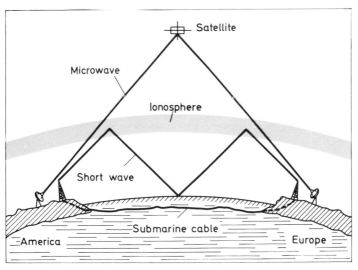

Fig. 1   Shortwave radio, submarine cable and satellite radio used for inter-continental communication

Table 1   Wave-(frequency-) ranges

Wavebands	Frequency ranges	Abbreviations
Very low frequency	10 to 100 kHz	VLF
Low frequency	100 to 300 kHz	LF
Medium frequency	300 to 1500 kHz	MF
High frequency	1.5 to 30 MHz	HF
Very high frequency	30 to 300 MHz	VHF
Ultra high frequency	300 to 3000 MHz	UHF
Super high frequency	3 to 30 GHz	SHF
Extremely high frequency	30 to 300 GHz	EHF

In this book the frequency ranges

    10 to 150 kHz (longwave range) and
    1.5 to 30 MHz (shortwave range)

will be discussed. Although the longwave and shortwave ranges account only for a small sector within the overall range of electromagnetic waves (Fig. 2), differences exist with respect to wave propagation even within the narrow shortwave range, and this applies in particular to sky-wave propagation. What makes the shortwave range attractive for message transmission is the fact that great distances can be spanned at moderate equipment cost and with comparatively low transmitting power (approx. from 100 W up to 1 kW).

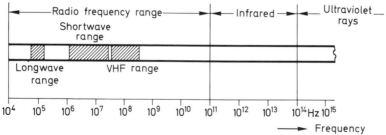

Fig. 2   Subdivision of wave ranges

1.2 Wave Propagation

The earth's atmosphere may be divided into several characteristic layers
(Fig. 3). The sphere in which we live, the **troposphere,** extends up to a height
of about 8 km near the poles and to about 20 km in the equatorial regions.
Still higher up is the **stratosphere** with its upper limit at about 30 km. This is
followed by the **ionosphere,** reaching up to about 400 km, and the
exosphere which merges into interplanetary space.
The electromagnetic waves radiated by a radio transmitter and propagating
close to the surface will be attenuated under the influence of the earth's
surface. In the shortwave range, these **ground waves** are subject to a comparatively heavy attenuation (Fig. 4) so that the transmission range is very

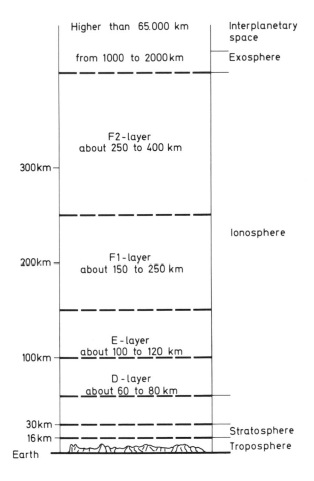

Fig. 3 Composition of
atmosphere

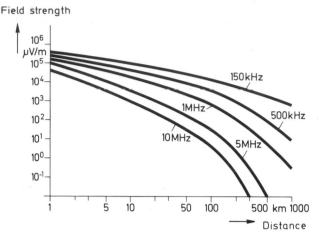

Fig. 4 Ground-wave propagation over land (transmitting power 1 kW, short vertical antenna)

small. It decreases with the frequency. Over land, distances of up to about 100 km can be spanned, and of about 1000 km over sea water. The ground wave is therefore only of limited use for shortwave circuits. If the ground wave is to be used for spanning greater distances, very low frequencies, e.g. 20 kHz must be selected. However, a very high transmitting power (about 1000 kW) and a costly antenna system will then be required. To avoid this heavy investment, commercial radio services employ the shortwave range for long-distance transmission. The longwave range between 50 and 150 kHz permits but a few transmitters to be operated. The difficulty of finding unassigned frequencies in this range compelled the system operators to use shortwave frequencies almost exclusively for new radio circuits. The mediumwave range is practically the domain of the entertainment sector.

1.3 Influence of Ionosphere on Wave Propagation

For spanning great distances in continental and intercontinental radio service only the **sky-waves** can be considered as useful means of propagation. If radiated under a favourable angle, these waves are reflected in the ionosphere back to the surface of the earth.

The gas atoms in these layers of the ionosphere, which are so important for the propagation of electromagnetic waves, are partly split up by solar

14

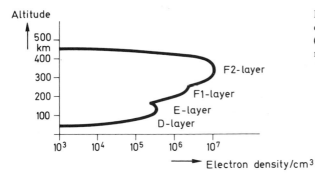

Altitude

Fig. 5 Electron density of the individual layers (by day, high sunspot number)

F2-layer
F1-layer
E-layer
D-layer

→ Electron density/cm³

radiation into positively charged ions and negatively charged electrons. The free and mobile electrons as well as the relatively heavy ions are responsible for the electrical conductivity which, however, varies greatly with the height above ground.

The ionosphere is made up of essentially four different layers located at different altitudes and having different thicknesses (Fig. 5). These layers, where reflection may occur, are referred to as **D-layer, E-layer, F1-layer** and **F2-layer**. The lowest or D-layer attenuates the waves penetrating it during the daytime. The attenuation introduced by this layer increases as the frequency decreases. For radio traffic it is the F2-layer which assumes greatest significance because it permits greater distances to be spanned than the E-layer.

Altitude and electron density on these layers are subject to continual fluctuations and this results in changes in the reflection and penetration characteristics. They depend essentially on the time of the day and the year, the number of sun spots, the geographic position and, last but not least, on the frequency used and on the angle of incidence. What has been said above applies to the E-layer with sufficient accuracy and normally also to the F1-layer. In the case of the F2-layer, however, deviations are frequently noticed which are due to phenomena like sunspot activity or magnetic storms.*

1.4 Refraction in the Ionosphere

When an electromagnetic wave penetrates into an ionized layer, the free electrons are excited and begin to oscillate. In so doing they absorb energy from the wave but, due to these oscillations, they in turn act as radiators

---

* The sunspot cycle has a period of about 11 years.

15

and release the absorbed energy again. A phase displacement exists, however, between the primary field of the exciting wave and the secondary field of the oscillating particles. This is the reason for the refraction of the waves, as the resulting field has changed with respect to the original field. The refraction increases with increasing electron concentration.

1.5 Attenuation of Waves in the Ionosphere

Electron motion is disturbed by collision with other particles. The energy of the incident wave is thereby reduced; in other words, the wave is attenuated. The lower the ionized layers, the more collisions occur, with a resultant increase in attenuation. The shortwaves pass through the D-layer but suffer their heaviest attenuation in this layer as well as in the E-layer. Thus, if a wave does not encounter low layers but is reflected in the F2-region, the reflection loss is very low. Multiple reflection from the F2-layer and the earth's surface permits distances to be covered which equal the circumference of the earth. To keep reflection losses down, the radiation angle of the transmitter antenna should be chosen so that the receiving station can be reached with a minimum number of steps. For long-range communication the angle of radiation will be rather small whereas a larger angle will be of advantage for shorter distances. Since the layers form under the influence of the sun's rays, their electron concentration depends on the time of the day and of the year. The E-layer is a day layer. It is comparatively stable. The F2-layer exists also at night but is subject to strong fluctuations. Table 2 shows the times when the individual layers are present. Figure 6 depicts wave propagation and the path of the waves in the ionosphere by day and by night. It can be clearly seen that the dead zone is far larger during the night.

Table 2 Contribution of the individual layers to shortwave propagation

Name of layer	Time of occurrence	Average altitude in km
D	By day, dependent on position of sun	60 to 80
E	By day, dependent on position of sun	100 to 120
F1	By day	approx. 200
F2	By day and by night	250 to 400

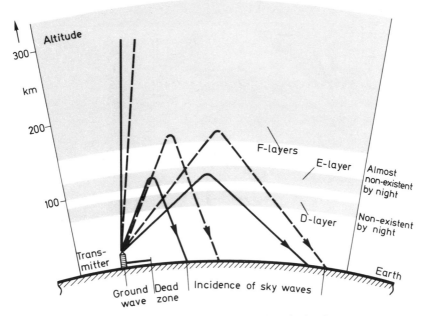

Fig. 6 Wave propagation and path of waves in ionosphere in the shortwave range between 1.5 and 30 MHz by day — and by night ----

1.6 Reflection in the Ionosphere

Incidence of waves at an angle causes a refraction in the ionosphere accompanied by a reflection. The electron density of the layers increases with the altitude, reaches a maximum, and decreases again. The refraction index of the layers follows the same law. Depending on whether the refraction index increases or decreases in the direction of the waves, the latter are refracted away from or towards the normal. The larger the angle of entrance into the layer, the more the waves must be bent if they are to return to the earth. Since the refraction index depends not only on the properties of the layer but also on the frequency, the diagram applies only to a specific frequency. Waves with a very small angle of entrance (1, 2, 3 in Fig. 7) pass through the ionosphere and are lost for radio traffic on earth. Waves 4 and 5 almost tangentially reach the central portion of the layer where there is no appreciable change in the refraction index. The curvature of the waves is therefore rather flat and the waves are deflected downwards only after a very long path. Under the influence of the increasing refraction

17

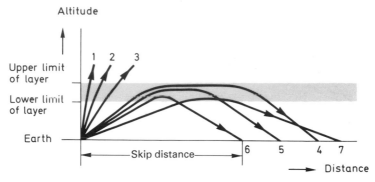

Fig. 7 Path of waves in ionosphere

index the wave will finally leave the layer and return to earth. In the case of wave 7 the change in the refraction index in the lower half of the layer is such that the wave is deflected and returns to earth almost at the angle of its entrance. The point where wave 6 impinges on earth is the one closest to the transmitter. Between this point and the location of the transmitter there is a dead zone which is characteristic of shortwave operation. It is practically impossible to receive in Germany the signals beamed from a German shortwave radio transmitter to South America. The distance between radio transmitter and point of impingement of wave 6 is the **skip distance.** In the case of the F2-layer (maximum height about 400 km) the maximum skip distance is 3500 km. For the E-layer (about 110 km high) it is about 2000 km.

1.7 Limits of Reflection in the Ionosphere

The frequency, that will just permit reflection, is called the critical frequency. This frequency varies with the degree of ionization of the layers. Above this limit no reflection takes place. The waves pass through the layers and reach interstellar space. This situation cannot be remedied by any increase in transmitting power.

As mentioned in Section 1.6, the refraction index is dependent on the frequency. The higher the frequency, the lower the refraction. Figure 8 shows the waves and the skip distances for four frequencies. As can be seen, the skip distance decreases with the frequency and the angle of incidence. Considering the fact that the sky-wave can only be received in the region beginning beyond the skip distance, a **maximum usable**

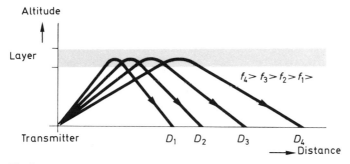

Fig. 8 Skip distance for four frequencies

frequency (MUF) exists for spanning any given distance. Using this frequency, the skip distance just equals the transmission range to be covered.

The lower limit of the usable frequency range is the **lowest usable frequency** (LUF) which is determined by the daytime attenuation in the D-layer. The attenuation is essentially a function of solar radiation.

Over short radio links, e.g. in the national networks with distances up to about 2500 km between transmitting and receiving points, uninterrupted traffic will often be possible with one frequency only or, if not, with two. For

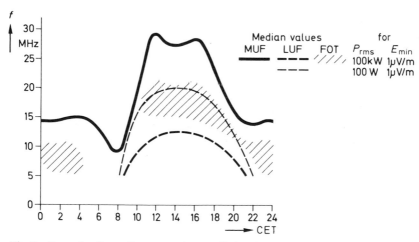

Fig. 9 Example of a radio propagation prediction for the route Hamburg–Rio de Janeiro for the month of January 1963, $R_{12} = 24$

19

overseas communications, however, several frequencies, depending on the distance and the time-zone difference between transmitting and receiving station, must be made available for sequential selection to ensure round-the-clock operation.

Figure 9 shows the MUF and LUF curves valid for the connection between Hamburg and Rio de Janeiro in January 1963. As can be seen, only the frequencies located between these curves can be used for operation at the indicated times. It should be noted that the usefulness of the ionosphere for message transmission gradually decreases in the direction of the MUF and the LUF. The **most favourable range** (FOT) will therefore be found between LUF and MUF, to be more precise, at about 15% below the MUF. Between this value and the MUF the field strength decreases by about 10 to 20 dB.

1.8 Delay Time Fluctuations on Shortwave Radio Circuits

The influences during wave propagation over shortwave links can cause delay time fluctuations in the transmitted signals. The bit length may therefore not fall below a specific value so that the signals can still be demodulated at the receiving end. In order to determine the magnitude of these delay time fluctuations, a number of measurements has been carried out on various radio paths. The results provide information concerning the

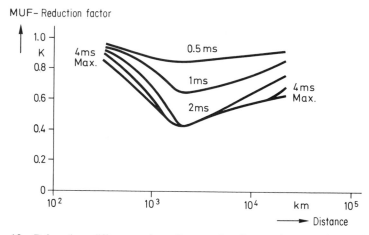

Fig. 10 Delay time differences depending on the distance between the sending and receiving stations and the MUF reduction factor k (after Salaman, R.K.: A new ionospheric multipath reduction factor. *IRE Trans. Comm. Syst.*, June 1962)

maximum telegraph rate at which binary signals can be transmitted on radio paths. Figure 10 illustrates the delay time fluctuations occurring on a radio path depending on the distance between the sending and receiving stations and on the amount by which the MUF is fallen below. This figure shows the relation between the MUF reduction factor k and the distance, assuming constant delay time differences:

$$k = \frac{f_b}{\mathrm{MUF}}$$

It can be seen that the closer the operating frequency f_b of the radio path is to the MUF, the smaller are the delay time fluctuations. Pronounced minima are found at a distance of approx. 2000 km between the sending and receiving radio stations. It can be seen here that the delay time fluctuation is approx. 0.5 ms when the operating frequency is approx. 15% below the MUF. If the deviation from the MUF is 50%, the delay time fluctuation reaches a value of 1.5 ms. Values for the maximum delay time fluctuations which can be expected on shortwave radio circuits are shown in Fig. 11. When the distance between the sending and receiving radio stations is over 2000 km, the maximum delay time fluctuation can be as much as 3 ms, whilst when the radio path is over 1000 km long a value of 5 ms is possible. The figures given are mean values which were determined from measurements made over a long period. The numerous influences which may cause the formation of possible reflection paths make a

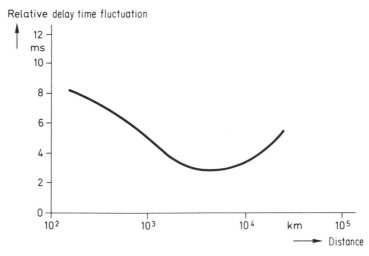

Fig. 11   Maximum delay time differences to be expected on shortwave radio circuits as a function of the distance (see also Fig. 10)

theoretical or even computed forecast for shortwave radio circuits difficult, and consequently one cannot go without measurements of the propagation conditions on each particular path.

1.9 Characteristics of the Radio Path and Disturbances

In Sections 1.5 to 1.7 the fact was mentioned that the field strength produced at the receiving station by the sky-wave is subject to perturbations during the propagation of the electromagnetic waves. The continual fluctuations of the ionosphere as regards structure, altitude, and ionic density, lead to a corresponding change in reflection conditions which in turn will affect shortwave radio communication.

Radio transmission may be impaired by

> fading,
> atmospherics,
> interference caused by electrical apparatus,
> interference from other radio transmitters,
> and white noise.

A knowledge of the disturbances affecting communication over shortwave links is important for the choice of suitable countermeasures intended to permit message transmission with a minimum of trouble.

The various phenomena will be dealt with in more detail in the following.

1.9.1 Fading on Shortwave Radio Links

As is known, space wave propagation with the aid of the ionosphere is not constant but subject to continual fluctuations because reflection conditions and, consequently, the field strength at the receiving station are continually changing. The result is a corresponding increase and decrease of the input voltage at the radio receiver. This phenomenon is called **fading**.

Depending on the causes, a distinction is made between

> absorption fading,
> interference fading, and
> polarization fading.

Variations in ionization result in **absorption fading.** This type of fading occurs at random intervals and the entire frequency band of a shortwave circuit is affected to the same extent. The **Mögel-Dellinger effect** which occurs occasionally for periods lasting from minutes to several hours is another phenomenon coming into this category. It is responsible for the sudden and simultaneous breakdown of nearly all shortwave connections in the daylight hemisphere. The connections in darkness, however, remain nearly unaffected.

These disturbances last from several minutes to an hour and when they have decayed the field strength will, in most cases, rapidly reach its original value. Many observations have led to the conclusion that Mögel-Dellinger perturbations are somehow connected with the appearance of large sunspots. The ultraviolet rays of the eruptions reach our planet with the velocity of light and these rays entail an increased generation of electrons in the lower layers. This causes the attenuation of the D-layer to reach a point where radio service is interrupted.

Interference fading, also known as **selective fading,** is due to the propagation of space waves over several paths between transmitter and receiver so that these waves arrive with a phase displacement at the receiving station.

If the phase displacement between these waves is zero, their amplitudes will be cumulative. A phase difference of 180°, on the other hand, results in complete cancellation or amplification, assuming equal amplitudes. If both the amplitude and the phase positions differ, the waves are only partly cancelled. This results in an irregular and transient increase and decrease of the field strength at the receiver. In contrast to absorption fading, these perturbations simultaneously affect only individual frequencies or a small band of frequencies extending over several hundred Hz and they progress slowly and irregularly through the entire frequency range. The other frequencies within the frequency band have their amplitudes changed to a greater or smaller degree. When digital signals are transmitted over narrow-band channels, parts of the message will be mutilated as a result of selective fading.

On the receive side selective fading thus causes the amplitude of the signal in the receive channel to change, as illustrated in Fig. 13. In order to be able to observe the behaviour of selective fading over a frequency band of about 2000 Hz, a voice band about 2700 Hz wide may be loaded with several closely neighbouring channels and a known signal transmitted. Figure 12 shows the received signal in this case. Here, selective fading in irregular sequence and of irregular width reduces the level, depending on the frequency and time. The intensity scale indicates the level from 0 dBm (black) to −48 dBm (white). Wandering of the fading may be clearly seen

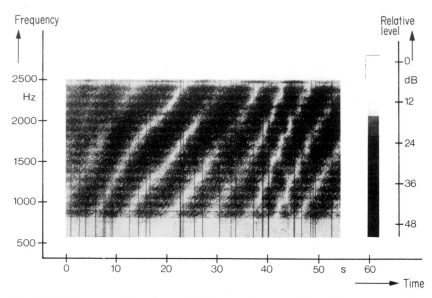

Fig. 12 Fading pattern dependent on the time and frequency, distance between the sending and receiving stations 1200 km (*IEEE Trans. Comm.*, Vol. COM-26, No. 6, June 1978)

here. Figure 12 illustrates one of the possible main groups of selective fading.

Besides the fading pattern shown, there are others which are often far more complicated.

Polarization fading is caused, in a similar way to interference fading, by double refraction of an initially linearly polarized wave in the ionosphere. The resulting wave has experienced a rotation of the plane of polarization. Polarization fading takes on the form of field strength variations occurring at regular intervals because the receiving antenna only selects a fixed polarization in most cases.

Fading is determined by three parameters; **depth, rate,** and **duration** (Fig. 13).

Variations in field strength (depth of fading) frequently reach a value of 40 dB — occasionally even 80 dB.

The fading rate ranges between 10 and 20 fades per minute according to a great number of statistical measurements. The frequency distribution of the duration of fading for a radio link is shown in Fig. 18.

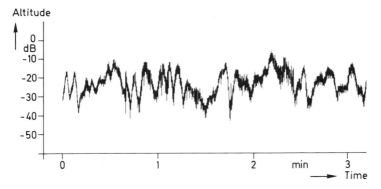

Fig. 13 Fluctuations of the IF output level of a shortwave radio receiver as a result of fading

As far as the duration is concerned, a distinction must be made between two separate phenomena: a duration of about 4 to 20 ms with rapid interference fading and of about 200 to 4000 ms with slow absorption fading.

The rate of variation of the field strength averages 4 dB/s as experience has shown. Peak values, however, may exceed 100 dB/s. The automatic gain control capability of modern radio receivers is such that variations of the input voltage from 80 to 100 dB are regulated down to $\pm$ 2 dB at a rate of up to 100 dB/s. Level variations are not in all cases fully compensated by the shortwave receiver. This applies in particular to broad-band transmission systems. To balance the remaining portion which may range between 20 and 40 dB, the telegraph demodulators following the radio receiver must be equipped with limiters.*

1.9.2 Interference affecting Shortwave Radio Reception

In addition to propagation conditions, the local noise level is of decisive importance for signal reception over shortwave paths. Where disturbances of different origin occur simultaneously, the field strength required for proper reception is determined by the strongest interference component. Interference phenomena may differ greatly as regards both type and amplitude and these depend strongly on local, diurnal and seasonal conditions and variations.

---

* The decrease in field strength (depth of fading) can be measured. To be able to measure frequency and duration of fadings, a threshold must be fixed for the level decrease (relative to the nominal level). Fading is then defined by the transgression of this threshold.

The most important sources of interference are:

Atmospherics

Atmospheric noise includes electrical discharges (lightning) as well as **white noise** (energy uniformly distributed over the entire frequency band). Electric discharges manifest themselves chiefly in the form of evenly distributed frequency spectra with occasional high-voltage peaks which are picked up by a radio receiver depending on its IIF or IF bandwidth and which reduce the signal-to-noise ratio. What is heard in the radio receiver is a crackling, frying and hissing noise. These perturbations are relatively frequent, in contrast to those which appear at regular intervals, i.e. annually or at intervals of several years.

The individual aperiodic voltage peaks occurring during thunderstorms as the result of electric discharges have a broad HF spectrum with the energy quickly decreasing towards higher frequencies. The frequency spectrum of the atmospherics depends essentially on the propagation conditions. Bandwidth and intensity of this spectrum vary considerably with the time of the day, the time of the year and the receive frequency.

The frequency of the **cosmic noise,** which may also interfere with signal transmission, extends from about 20 to 1000 MHz.

Man-made noise

Radio reception is also impaired by electrical machinery and appliances which are not adequately r.f.-suppressed (keeping to a certain minimum noise level) and by unwanted r.f. radiation from medical electronic equipment. Man-made noise may be caused by arcing of collectors and switches, by ignition sparks of car engines, etc. Its amplitude depends strongly on local conditions, also on the power supply, since a considerable portion of the noise energy is transmitted by way of the commercial power network. In noisy areas the level of the atmospherics appearing during the day in the shortwave range is rather low so that man-made noise predominates.

Interference from other radio services

Important sources of interference are other radio services, for example, if a number of independent radio transmitters are using the same frequencies or are operating on adjacent frequencies with insufficient channel spacing. Where a number of co-located transmitters are operating with different frequencies, interference may be caused by cross-modulation. A CCIR recommendation [3] has therefore been issued to fix the permissible limits of the frequency spectrum and of the bandwidth of radio transmitters.

1.9.3 Types and Effects of Interference

If the above-mentioned interference comes within the frequency range of the useful signal, an appreciable reduction in the quality of transmission will be the result. The influence of interference increases as the difference in the amplitude between the noise signal and the useful signal decreases. The measure for this is the signal-to-noise ratio which is expressed in dB.

$$\frac{S}{N} = \frac{\text{Signal}}{\text{Noise}}$$

With a very low incoming signal level, caused by fading, the interference may make itself felt to such an extent that undisturbed message transmission is no longer possible.

Interference may be classified as follows:

Noise energy which is almost uniformly distributed over a wide frequency band and over the time (white noise),

Crackling, in most cases only of short duration, but affecting broad frequency bands,

Energy from other radio transmitters and discrete interferers which hits a frequency band selectively (intentionally or unintentionally) and which may be present continuously.

Figure 14 shows the assignment of frequencies as a function of time with different types of interference. In practice a mixture of these types of interference will occur.

Careful, time-consuming measurements have been carried out over a period of many years on shortwave circuits of different lengths with special emphasis on evaluating transmission errors caused by fading and other interference as to their number and nature. It has been found that the mutilation of message signals is essentially due to fading and interference. In the recent past, errors in a test message sent over a shortwave circuit have been recorded on perforated tape to investigate the nature of the errors (Fig. 15) with a view to determining the distribution and duration of dis-

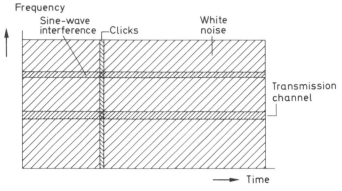

Fig. 14   Various types of interference on shortwave radio
links as a function of frequency and time

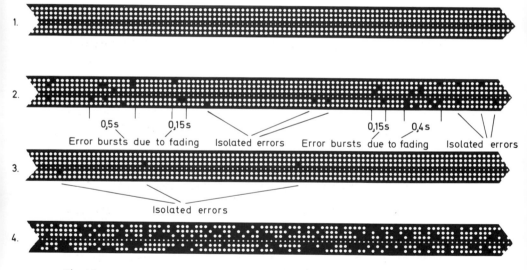

Fig. 15
Example of an
error structure of
a shortwave link.
(Each hole corresponds
to a correctly
transmitted bit.
A missing hole implies a
bit falsified *en route*)

1. Error-free operation
2. Errors caused by fading
3. Isolated errors caused by clicks
4. Interference caused by high-power radio
   transmitters
Length of radio link 120 km
Operating frequency 4.8 MHz
Telegraph speed 100 bits/s (1 hole = 10 ms)
Frequency shift ±42.5 Hz

28

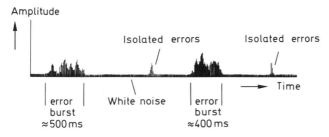

Fig. 16 Randomly distributed isolated errors and burst errors

turbances.* The structure of the errors can be readily recognized. Almost every second bit is falsified within the error bursts occurring during fading periods.

Interposed between the error bursts are periods of error-free transmission. During fading periods the useful signal energy at the receiver has dropped to the point where the energy picked up by the radio receiver is practically only noise energy. In addition to these error bursts, the punched tapes reveal isolated errors which are presumably due to clicks. If the interferer was another radio transmitter, the major portion of the original information will have been lost.

As regards frequency and distribution in time, disturbance affects message transmission differently, depending on its source. A distinction is made between two basic forms, **random errors** and **burst errors** (Fig. 16).

Random errors occur, for instance, when the signal is drowned out by white noise. The probability of a bit being mutilated is the same for each bit of the message, independent of whether adjacent bits are mutilated or not.

Error bursts are the result of fading. Generally speaking, they are characterized by two parameters: by their frequency as compared to the total number of signals transmitted, and by their duration. There exists no regularity as regards the frequency and the duration of the bursts so that the original bits are replaced by an arbitrary sequence of bits as long as the burst lasts. An error burst within a serial, isochronous sequence of bits may be defined as a range having the following characteristics: a burst invariably begins and ends with a falsified bit. This range comprises a number of

---

* Transmission errors on shortwave radio links are recorded by having the originating station transmit a test message whose consecutive bits all have equal length. At the receiving station, where the text is known, the incoming signals are compared bit by bit with the expected signals. To obtain a record of the bit error distribution, a hole is punched into the tape for each bit that was correctly received. A missing hole indicates a falsified bit.

29

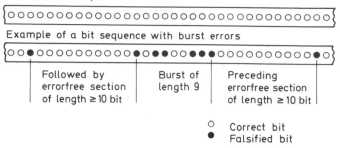

Fig. 17   Schematic representation of burst definition

falsified bits, hence a certain bit error density. The bit error density is the ratio of the number of the falsified bits to the total number of bits transmitted within this range. Each burst is enclosed between two intervals of a bit error density lower than that of the burst period. These intervals begin and end with a correct bit. The definition of a burst is given schematically in Fig. 17.

Most of the disturbances occurring in shortwave operation are bursts. These are caused by fading.

Figure 18 shows the frequency distribution of the duration of fading on an experimental circuit, revealing the frequency of possible burst durations. At a frequency of 10% the duration ranges between 40 ms and 500 ms.

1.9.4 Transmission Range of a Radio Link

The transmission range of a radio link depends on the transmitting power, the frequency employed, the gain of the transmitting and receiving antenna (directivity), the noise level at the receiving end and the sensitivity of the radio receiver in the receiving installation. To extend the range of a transmission system to its very limit, due consideration should be given to the gain of the transmitting antenna and of the receiving antenna, to a high sensitivity of the receiver, and to the transmission conditions and the noise level at the receiving station.

1.9.5 Level Conditions on the Radio Path

Apart from the capabilities of the radio equipment and the associated antennae, the quality of a radio circuit depends on the modulation method

30

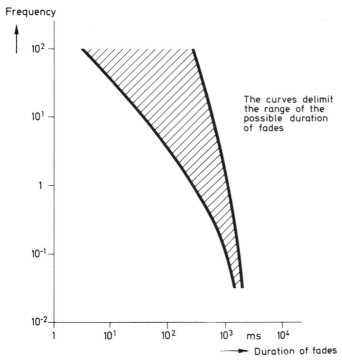

Fig. 18 Frequency distribution of duration of fading on an experimental shortwave link (Frankfurt/Main–New York–Frankfurt/Main) (H. Retting and K. Vogt, *NTZ*, **17**, 2 (1964))

employed for digital message transmission. An essential factor is, however, the minimum receive level of a radio station. Figure 19 shows a level diagram of a radio link. Where great distances have to be spanned, the level of the transmitted signal decreases considerably along the radio path due to attenuation losses. Furthermore, the useful signal level may be reduced at the input of the radio receiver as a result of fading. The basic noise level must also be taken into consideration. For satisfactory operation, the transmitting power at the transmitting station must be high enough to ensure an adequate incoming signal level (at the receiving station) for a certain error rate in spite of attenuation, fading, noise level and interfering radio transmitters. In this case any further increase in the transmitting power will not bring about an appreciable improvement in the quality of the radio link, and better results can then only be achieved by having recourse to diversity or message protection methods.

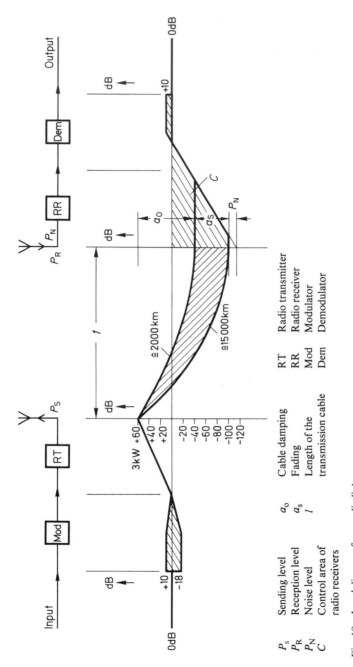

Fig. 19 Level diagram for a radio link

P_S Sending level a_O Cable damping   RT   Radio transmitter
P_R Reception level a_S Fading   RR   Radio receiver
P_N Noise level l Length of the   Mod   Modulator
C Control area of   transmission cable   Dem   Demodulator
                  radio receivers

32

The values given in Fig. 19 are values measured in a practical case. The attenuation between transmitting station and receiving station is assumed to be about 120 dB. The depth of fading ranges, on an average, between 30 and 50 dB. The receiving equipment amplifies the signal and compensates for level variations so as to obtain an average value.

1.10 Propagation Conditions in the Longwave Range

Telegraph transmission in the longwave range between 50 and 150 kHz is employed essentially in continental traffic. The receiving conditions may be expected to be more stable than in the shortwave range both in the daytime and at night. In view of the high transmission reliability of longwave radio circuits the longwave range has been crowded with the transmitters of news agencies and meteorological services. The Central European area is served by medium-power transmitters.

The attenuation of the ground wave is very low in the longwave range and this results in a large fade-free zone. On the other hand, the sky-wave is heavily attenuated during the day by the D-layer. Local fading is therefore likely to occur at intervals of several hundred kilometres from the transmitter as a result of interference between ground wave and sky-wave. Within this range the received signals have nearly the same level. At a greater distance from the transmitter, e.g. if the distance exceeds 600 km, signal reception is determined by the ground wave during the day and by the sky-wave at night. By night, fade periods lasting several minutes may occur, which, because of their low depth, do not impair signal reception. At distances of about 100 km from the transmitter, fading occurring at sunrise and sunset may temporarily result in cancellation of the sky- and ground waves due to interference in the transition period between day and night fields. For Central European conditions the above distances are only guiding values because wave propagation depends strongly on ground conditions. The transmission range obtained with the ground wave increases appreciably over water. In Fig. 20 the field strength of the ground wave is shown as a function of the distance between transmitter and receiver.

In longwave communication, atmospherics make themselves particularly felt during the thunderstorm-prone months. Also a strong increase in radio interference, caused by electrical equipment, is experienced in this range. To ensure satisfactory reception the signal field must be stronger than in the shortwave range.

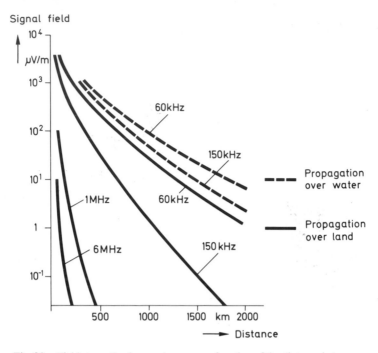

Fig. 20   Field strength of ground wave as a function of the distance between transmitter and receiver (referred to a transmitting power of 1 kW)

1.11 Conclusions

Transcontinental radio traffic over shortwave links is made possible in most cases by reflection of the sky-wave from the conducting layers of the ionosphere. Radio reception is impaired by different phenomena and, in the worst case, may be completely marred. The complex inter-relationships have been dealt with in detail in the individual paragraphs of this chapter.

Thus, for planning radio links, it is important to know the propagation conditions of the ionosphere with a view to the measures to be taken at the transmitting and at the receiving end, the transmission and modulation methods to be adopted, and the requirements to be imposed on the radio transmitters and receivers and on their antennae. The receiving levels and the noise levels must be estimated so that the requirements with respect to signal-to-noise level and fading performance of the radio receivers and demodulators can be taken into account.

2 Message Structure

2.1 Digital Signal

In telegraph and data communication, information is transmitted in the form of characters, namely letters, figures and symbols. The information is represented by binary signals, which are characterized by different states. When considered electrically, these signals correspond to, for example, the states 'current' and 'no current' or '1' (H) and '0' (L), or in accordance with CCITT recommendations 'stop polarity Z' and 'start polarity A'. In carrier frequency communication the binary signal corresponds to the states 'tone ON' and 'tone OFF' in the case of amplitude modulation, to two frequencies in the case of frequency modulation and to two phases of a carrier wave in the case of phase modulation. The binary signal is thus characterized by two states (Table 3).

In digital message transmission over radio circuits, the signal elements of the characters are transmitted in turn (serially). The binary states and their

Table 3   Binary signal states

	State during the:—	
	Start pulse	Stop pulse
Designation as per CCITT	start polarity	stop polarity
Characteristic state as per CCITT	A	Z
Earlier designation	Z	T
English designation	space	mark
Binary digit	0 (L)	1 (H)
Neutral current keying (closed-circuit operation)	no current	current
Polar current keying	negative current or positive current	positive current or negative current
Amplitude modulation	no tone	tone
Frequency modulation	higher frequency	lower frequency
Punched tape	no hole (O)	hole (●)
Differential phase modulation	phase shift	no phase shift
Facsimile transmission	white	black

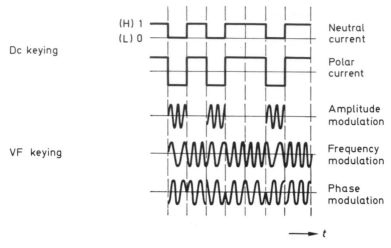

Dc keying

(H) 1
(L) 0

Neutral current

Polar current

VF keying

Amplitude modulation

Frequency modulation

Phase modulation

t

Fig. 21   Assignment of binary conditions to modulation methods

assignment to the keying and modulation methods used in telegraphy are shown in Fig. 21. The messages can be transmitted using a.c. or d.c.

In telegraph and data communication, a distinction is made – depending on the type of transmission method employed – between **start–stop systems** and **synchronous systems.** In the case of start–stop systems, the character code group consists of the start pulse, the signal elements and at least one stop pulse (Fig. 22). Synchronism between the transmitting and the receiving stations is established for the duration of one character through recognition of the start pulse.

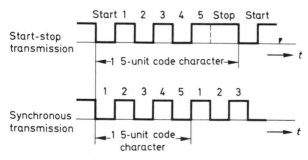

Start-stop transmission

Start 1   2   3   4   5   Stop   Start

←1 5-unit code character→

t

Synchronous transmission

1   2   3   4   5   1   2   3

←1 5-unit code character→

t

Fig. 22   5-unit code characters with start-stop and synchronous transmission

36

In start—stop systems, the start pulse and the stop pulse have the task of synchronizing the transmitter and receiver of the teleprinters for the period of one telegraph character. The duration of the stop pulse is usually of unit length or 1.5-times unit length, to ensure that the teleprinter receiver camshaft stops properly during tape transmissions.

In the case of synchronous systems, the transmitter and the receiver are always in synchronism. In contrast to start—stop systems, only pulses of uniform length are used (isochronous pulse sequence).

For transmission, the three terms **modulation rate, bit rate** and **character rate** are of importance.

2.1.1 Modulation Rate

The modulation rate v_s — also called telegraph speed — is the reciprocal of the duration of a pulse T_o measured in seconds. It also specifies the maximum number of pulses which can be transmitted in a second. On reducing the pulse length, the modulation rate increases:

$$v_s = \frac{1}{T_o}$$

The unit of modulation rate is the baud (1 baud = 1 pulse/s).

Example

For $v_s = 100$ bauds the pulse length $T_o = 10$ ms.

Table 4 lists the current trend in modulation rates for transmission systems using shortwave radio links.

Table 4   Modulation rate and bit rate of different terminal equipments and transmission systems

Teleprinter	45.45	50	75	100	bauds
Data transmission	200	600	1200	2400	bits/s
Time-division multiplex system	2-channel 4-channel		96 192		bauds bauds
Facsimile telegraphy			to 3600		bauds

2.1.2 Bit Rate

The bit rate is the product of the modulation rate (symbol rate) v_s and the number of bits transmitted per symbol.

In synchronous transmission systems, the maximum bit rate is numerically equal to the modulation rate. If, for example, in the case of binary single-channel serial transmission over a radio link, the modulation rate is 100 bauds, a maximum of 100 bits per second flow through the circuit. The bit rate is therefore 100 bits/second.

Under the same conditions, the bit rate is lower for start–stop systems. A 5-unit code character contains 5 information bits. The start and stop pulses contain no information. With a stop pulse of 1.5-times unit length and 20-ms signal elements, the maximum bit rate v_B is:

$$v_B = \frac{\text{Bits per character}}{\text{Transmission time per character}} = \frac{5 \text{ bits}}{0.15 \text{ s}} \approx 33\tfrac{1}{3} \text{ bits/s}$$

2.1.3 Character Rate

The character rate v_C is a measure of the performance of the data terminals or teleprinters. It indicates the number of characters which can be transmitted or received per unit of time.

Taking as an example a 50-baud teleprinter with each character consisting of a start pulse, 5 signal elements and a 1.5-times unit length stop pulse, the character rate v_C is:

$$v_C = \frac{1 \text{ character}}{\text{Transmission time per character}} = \frac{1 \text{ character}}{0.15 \text{ s}} = 6\tfrac{2}{3} \text{ characters/s}$$

or 400 characters/min.

2.2 Types of Distortion

When telegraph characters are transmitted over a radio link, which consists of radio transmitter, transmission path and radio receiver, the length of the individual pulses of a character can be changed. Such changes in length significantly affect the reliability with which the characters are correctly received. The term **signal element distortion** has been introduced as a measure of these deviations [7], this was at one time called telegraph distortion. The quality of the transmission is derived from the degree of signal element distortion. If certain distortion limits are exceeded, correct interpretation of the pulses on the receiving side can no longer be guaranteed.

A distinction is made between **isochronous distortion** and **start–stop distortion.** Distortion also appears in different forms, e.g. as bias and characteristic distortion.

2.2.1 Signal Element Sequences and Signal Element Distortion

In order to transmit letters and figures in telegraph and data communication systems, they are assigned characters consisting of certain signal element sequences. The transmitting and receiving teleprinters operate at the same modulation rate. The length of the signal elements is thus fixed. The modulation rate on radio telegraph links is usually 50 bauds, which corresponds to a signal element length of 20 ms. The transmitting teleprinter thus emits pulses in this case, which can assume one of two characteristic states, for a period of 20 ms with the associated characteristic state. The receiving teleprinter has the entire pulse duration at its disposal for pulse sampling, but makes do with a much shorter period, e.g. 10% of the pulse length. This sampling period is timed to occur at the centre of the received pulse. It is easy to understand that the characters are printed free of error, if the pulses are available during the sampling period with the correct polarity. The interval between the theoretical start of the pulse and the start of the sampling procedure by the teleprinter receiver constitutes a margin in which the start of the pulse, i.e. the pulse transition, may fall without causing transmission errors. The magnitude of the displacement of the pulse transition with respect to the nominal instant determines the quality of a digital transmission channel. The signal element distortion provides a constant measure of transmission quality. Transmission errors do not occur until the signal element distortion exceeds the reception margin. Figure 23 shows the effect which the signal element distortion has on transmission quality. The pulses are transmitted using a time base. At the receiving end they are sampled at their theoretical centres, referred to the leading edge. As long as the polarity is correct during sampling, the teleprinter supplies the characters free of error. However, if the signal element distortion is so large that it exceeds the reception margin and the pulse transition thus lies in the sampling interval, the teleprinter receiver interprets the mutilated incoming characters in the normal manner and prints a falsified message.

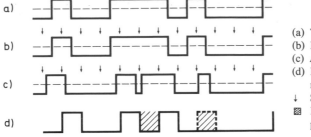

	(a) Transmitted function
	(b) Ideal received function
	(c) Actual received function
	(d) Interpretation in
	receiver
↓	Sampling in receiver
▨	Incorrectly interpreted
	pulses

Fig. 23   Errors in the transmission of telegraph and data signals

2.2.2 Individual Distortion

Individual distortion is the deviation of the pulse transition, i.e. the deviation of the characteristic instant from the theoretical nominal instant. Figure 24 shows an example of individual distortion. The instants t_1 to t_4 are the theoretical nominal instants. The first and last pulse transitions correspond to the theoretical instants and thus exhibit no distortion. The second pulse transition occurs after its nominal instant; it is distorted in a lagging sense.

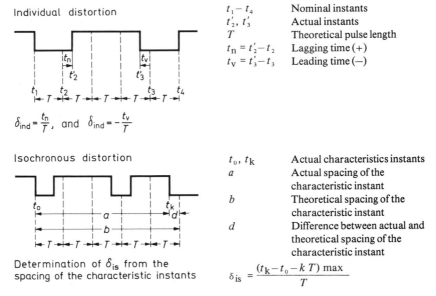

Individual distortion

$$\delta_{ind} = \frac{t_n}{T}, \text{ and } \delta_{ind} = -\frac{t_v}{T}$$

$t_1 - t_4$	Nominal instants
t'_2, t'_3	Actual instants
T	Theoretical pulse length
$t_n = t'_2 - t_2$	Lagging time $(+)$
$t_v = t'_3 - t_3$	Leading time $(-)$

Isochronous distortion

Determination of δ_{is} from the spacing of the characteristic instants

t_0, t_k	Actual characteristics instants
a	Actual spacing of the characteristic instant
b	Theoretical spacing of the characteristic instant
d	Difference between actual and theoretical spacing of the characteristic instant

$$\delta_{is} = \frac{(t_k - t_0 - kT)\,\text{max}}{T}$$

Fig. 24   Individual and isochronous distortion

40

The third pulse transition arrives before its nominal instant; it is distorted in a leading sense. Lagging distortion is designated by +, leading distortion by −.

The degree of individual distortion is

$$\delta_{ind} = \frac{t' - t}{T},$$

where t is the nominal instant of the pulse transition,
$\qquad t'$ is the actual instant of the pulse transition and
$\qquad T$ is the theoretical pulse length.

The distortion is specified as a percentage of the theoretical pulse length.

The individual distortion can be used as the theoretical foundation for the derivation of further distortion definitions. It is, however, problematical in practice to determine the individual distortion of each characteristic instant. In any case, it is usually impossible to determine with accuracy the reference point, i.e. the nominal instants. The exact position of the nominal instants is lost in transmission.

The individual distortion can be determined as follows. The transmitter of a data tester supplies the signal elements of a character using a nominal time base.

If the pulses at the output of the data tester are distorted, for example due to an inaccurately adjusted output circuit, the individual distortion of the output signals can be determined exactly by referring to the transmitter-internal nominal time base. Should there, however, be a transmission section between the signal to be measured and the nominal time base source, it is usually impossible to determine the individual distortion with accuracy. By taking special measures it is, however, possible to derive the nominal time base approximately from the incoming signals, so that the individual distortion can be measured with reduced accuracy even over transmission sections.

2.2.3 Isochronous Distortion

In synchronous transmission the isochronous distortion can be measured. Isochronous distortion (Fig. 24) is defined as the algebraic difference between the largest and the smallest value of the individual distortion. The instants of the nominal time base are thus not taken into consideration, i.e. the isochronous distortion is only referred to the nominal pulse length.

Degree of isochronous distortion

$$\delta_{is} = (\delta_{ind\ max} - \delta_{ind\ min}) = \frac{\Delta t_{max} - \Delta t_{min}}{T}$$

When the isochronous distortion is determined, it is therefore no longer necessary to refer the individual characteristic instants to the nominal instants, only the relative position of the characteristic instants with respect to each other is measured. It is thus possible to measure the isochronous distortion at any point on a transmission path. The measuring instrument itself must only compare the spacing of the characteristic instants of the measured signal with an internally generated nominal spacing.

The degree of isochronous distortion δ_{is} indicates the spread of the characteristic instants referred to the pulse length.

2.2.4 Start–stop Distortion

The teleprinters and the punched tape equipment use the start–stop method of operation. The effect of the signal element distortion on the reception of the characters is indicated by the start–stop distortion. The start–stop distortion refers, like the individual distortion, to nominal instants. However, the position of the nominal instants for the pulse transitions of the character is governed by the start pulse transition. Just as in the case of isochronous distortion, it is therefore possible to determine this distortion at any point on the transmission path. The start–stop distortion can, of course, only be measured in asynchronous transmission, i.e. start–stop transmission. Start–stop distortion also occurs in a leading (–) and a lagging (+) sense, depending on whether the characteristic instant is before or after the nominal instant.

The degree of start–stop distortion is

$$\delta_{st} = \frac{|t' - t|_{max}}{T} = \frac{|\Delta t|_{max}}{T}$$

2.2.5 Typical Forms of Distortion

It is frequently possible to draw conclusions as to the distortion cause from the typical forms of distortion. The most important will be discussed as follows:

▶ *Bias distortion*

With bias distortion all pulses of one characteristic state are increased in length, and all those of the other characteristic state are correspondingly reduced in length. This distortion is typical for maladjusted transmission circuits. In this respect, special mention should be made of telegraph relays, discriminators (for conversion of frequency modulated a.c. voltages into d.c. voltages) etc.

▶ *Characteristic distortion*

The characteristic distortion is usually caused by transients and is system-dependent. It is noticeable when the duration of the individual characteristic states varies greatly. If the two characteristic states are of the same duration, no characteristic distortion occurs. Such distortion is principally encountered where only a narrow bandwidth is available for transmission, i.e. on VF telegraph circuits in which signals are either transmitted at speeds considerably above the rated values, or with an excessive outgoing distortion.

2.3 Error Frequency

Errors occur during message transmission when the signal element distortion is so severe that it exceeds the receive margin. In order to record transmission errors, a standard test pattern is transmitted and this text is compared with the text actually printed out on the receiving side.

The quality of a radio link is also indicated by the error frequency (error rate). A distinction is made between bit error rate and character error rate.

The yardstick on radio telegraph links is usually the character error rate, the number of falsified characters being counted:

$$P_C = \frac{\text{Sum of characters falsified during transmission}}{\text{Number of characters transmitted}}$$

Example

A character error rate of 1×10^{-5} means that on the average one character is falsified per 100 000 transmitted characters. Table 5 shows the average character error rates of various transmission circuits.

43

Table 5 Character error rate for telegraph and data transmission over shortwave radio links compared with telecommunications networks (e.g. cable and radio relay circuits). The teleprinter was included purely for comparison purposes.

Type of message transmission	Average character error rate	Average message volume between two errors on fully typed pages (2500 characters per page)
Unprotected shortwave radio link	1×10^{-2} to 1×10^{-3}	2 to 20 lines
Shortwave radio link, data protected	1×10^{-5} to 1×10^{-6}	40 to 400 pages
Dedicated telegraph circuit, 50 bauds	1×10^{-5} to 1×10^{-6}	40 to 400 pages
Dedicated telephone circuit	0.5×10^{-4} to 1×10^{-5}	8 to 40 pages
Telex network, 50 bauds	3×10^{-5}	15 pages
Teleprinter	1×10^{-6}	400 pages

If a transmission channel is affected by interference, a '1' may appear after demodulation instead of '0' or vice versa. This is referred to as a bit error.

The bit error rate is analogously the sum of the bits falsified during the transmission referred to the number of bits transmitted:

$$P_B = \frac{\text{Sum of bits falsified during transmission}}{\text{Number of bits transmitted}}$$

For the objective assessment of a transmission system with regard to transmission quality, the theoretical or measured bit error rate P can be specified as a function of the normalized signal-to-noise ratio R [12]. In the case of white noise, it is relatively easy to measure the bit error rate as a function of the transmission bandwidth. Figure 25 shows the theoretical bit error rate as a function of the quantity R for an ideal frequency-modulated transmission system. In accordance with CCIR Report 195 (Geneva 1974) the curve was determined using the following designations:

v Modulation rate in bauds or signal flow in bits per second of the binary FM signal
B Bandwidth occupied by the signal in Hz
R Normalized signal-to-noise ratio

44

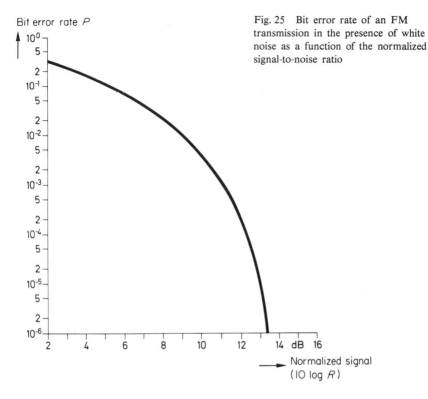

Bit error rate P

Normalized signal ($10 \log R$)

Fig. 25 Bit error rate of an FM transmission in the presence of white noise as a function of the normalized signal-to-noise ratio

The normalized signal-to-noise ratio is determined as follows:

$$R = \frac{\text{Energy per bit}}{\text{Noise power per Hz}} = \frac{\text{Signal power/modulation rate}}{\text{Noise power per Hz}}$$

For a system implemented in practice, the measured curves lie a few decibels further to the right, as shown in Fig. 47. The spacing from the ideal curve – measured in dB – is a yardstick of the quality of the transmission system.

The bit error rate of a transmission system due to noise can also be determined from the signal-to-noise ratio in the occupied transmission channel with frequency band B:

$$r = \frac{\text{Signal power}}{\text{Noise power in the occupied bandwidth}} = R \times \frac{v}{B}$$

At a certain modulation rate v, the quantity r may be increased by keeping

45

the bandwidth B as small as possible. For a bandwidth matched to the modulation rate, the optimum is when $v = B$. This corresponds to the results which concern the modulation index m suitable for binary frequency modulation. For $m = 2/\pi \approx 0.67$ to $m = 0.8$, a particularly compact frequency spectrum with a width approximately satisfying the condition $v = B$ is obtained for modulation with binary signals. In contrast to this, a slightly larger modulation index is required for transmission systems using frequency modulation, if – for all code combinations – the distortion of the signals is to be particularly low in the presence of noise.

The following example demonstrates how, on the basis of the previous considerations, the error rate of a matched transmission channel can be reduced to a minimum.

A 50-baud transmission channel, matched to the modulation rate, with minimum error frequency is to have the following characteristics:

> *Bandwidth* ± 25 Hz
>
> *Frequency shift* ± 20 Hz (for $m = 0.8$)

Inadequate frequency stability of earlier radio equipment made it necessary to use filters with an extremely great bandwidth and large frequency shifts of, for example, ± 400 Hz for a 50-baud message. On employing the above values with matched bandwidth, a gain can be achieved which is characterized by a 'bandwidth factor' of 20, that is an improvement of approximately 13 dB in the signal-to-noise ratio. The effect this has on the bit error rate is shown in Fig. 47. It is practicable to expand the bandwidth from 50 to 60 Hz on account of the frequency drift which can never be completely avoided. It is present to some extent even with modern equipment (approximately 5 Hz) and may also be caused by the transmission path.

2.4 Telegraph Alphabets

In the early stages of radio engineering messages were transmitted by simply cutting the carrier of a transmitter in and out. The Morse alphabet was used for this purpose. This alphabet still has its uses today in, for example, amateur radio, maritime radio and mobile radio communication.

The present day teleprinters use a five-digit binary code, which provides a total of $2^5 = 32$ characters (International Telegraph Alphabet ITA No. 2), as shown in Fig. 26.

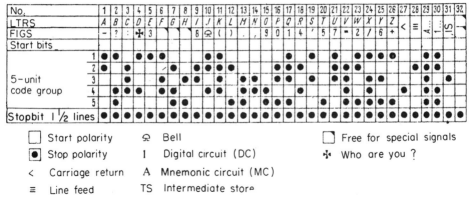

Start polarity   ♀ Bell   ☐ Free for special signals

● Stop polarity   I Digital circuit (DC)   ⚓ Who are you ?

< Carriage return   A Mnemonic circuit (MC)

≡ Line feed   TS Intermediate store

Fig. 26   Code table for the International Telegraph Alphabet ITA No. 2

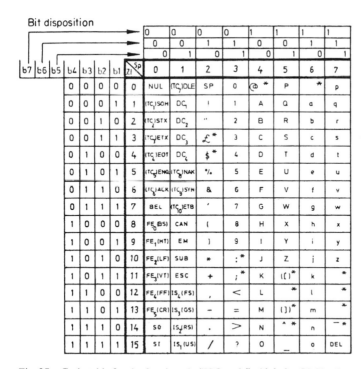

Fig. 27   Code table for the 8-unit code ('ISO code'), Alphabet IA No. 5

This is enough for the transmission of the letters, but to be able to transmit figures and the symbols customary in printing, two of the code combinations are used to switch the teleprinter from letters to figures and vice versa. By means of this shift, the original total of 32 characters can be nearly doubled. An adequate number of transmittable characters is therefore available for the transmission of most texts. Codes with a higher number of digits are used for message protection and in data transmission.

For data transmission, the 5-unit code is usually inadequate, as a number of code combinations are required for control purposes and the letters/figures shift is omitted. In accordance with CCITT recommendations an 8-unit code (Alphabet IA No. 5, formerly called 'ISO code') consisting of seven information bits and a parity bit is here employed (Fig. 27).

2.5 Facsimile Telegraphy

Facsimile methods are used for the transmission of weather charts, drawings, documents or handwritten notes. In the shortwave range, the transmission of weather charts accounts for the major portion of all facsimile traffic. The mode of operation of weather chart recorders is therefore dealt with in the following paragraphs.

The copy obtained at the receiving end should be as exact a replica as possible of the transmitted original. Facsimile units employ the photoelectric scanning process. The light levels are converted into the equivalent electrical signals which are transmitted over the radio channels to the receiving stations. This is a binary process with only two conditions, black or white, being transmitted. The subject copy is scanned line by line. Recording at the receiving end is also line by line. The quality of a facsimile transmission over radio links depends essentially on the bandwidth of the radio channels. Meteorological services prefer black and white transmission in addition to the transmission of gray tones (picture transmission).

Mode of operation

The operating principle of facsimile transmission is based on the decomposition of the subject copy into lines and the line-by-line scanning of this copy. The copy is wrapped around a picture drum (in the case here discussed). Other versions work without a drum.

A scanner moves axially in front of the rotating drum and directs a sharply focused beam from a light source onto the copy. It converts the black and

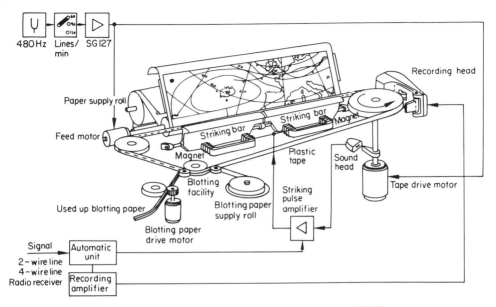

Fig. 28 Schematic representation of the operating principle of a HELLFAX met chart recorder

white values (brightness levels) into electrical values and transmits them in the form of binary signals. The synchronized recorder is designed to reproduce the 'original copy' line by line. The receive-side recorders are continuous paper recorders in most cases. Indirect recording, i.e. print transfer, is frequently employed. The incoming signals are first written on a continuous, revolving tape from which a complete line at a time is then transferred onto the actual recording paper. A blotter removes the residual ink from the plastic tape. Figure 28 is a schematic representation of the operating principle of the HELLFAX met chart recorder. Modern facsimile recorders use electrosensitive paper on which black traces appear at the points of current passage. Figure 29 shows the HELLFAX met chart recorder, a continuous-paper facsimile recorder employing electrosensitive roll paper.

The **drum speed** *n,* i.e. the number of the lines scanned per minute, has been standardized at 60, 90, 120, 180 and 240 rpm. A higher drum speed involves an increase in the message volume transmitted per time unit and in the bandwidth required for the transmission channel.

Fig. 29   The HELLFAX meteorological chart recorder

The **transmission time** t of a subject copy is derived from the number of the lines of the entire copy and from the number of the lines transmitted per minute. These transmission times are determined by the index of co-operation, the drum speed and the length of a chart.

The index of co-operation M is obtained as the product of drum diameter and the number of lines scanned per millimetre. For proper inter-operation the transmitting and the receiving facsimile equipment must employ the same index of co-operation, otherwise the received copies will be either smaller or larger. The index of co-operation stipulated by CCITT for met chart transmissions is either 288 or 576.

Synchronization is ensured by phasing the receiving equipment prior to transmission. This is accomplished by the transmitting station sending phasing pulses for a brief period. Fully automatic operation is also possible. In this case the transmitting station will control the recorders by transmitting special start and stop frequencies. Thus, the desired index of co-operation can be remotely adjusted from the transmitting station. The recorder may also be started and stopped by remote control whereas the speed setting is an automatic process.

Picture frequency and bandwidth of facsimile signal

If a copy is scanned with four lines per millimetre, the resulting line width is 1/4 mm. Since this resolution must be adhered to both in the feed and in the scanning directions, the scanner must be capable of evaluating an area of 1/4 mm × 1/4 mm. The maximum number of black-to-white changes that can be covered by such a piece of equipment is obtained when a checkerboard pattern with an edge length of 1/4 mm has to be scanned. In the case of a DIN-A4 copy this results in (210 × 4) × (298 × 4), i.e. in approximately 1 000 000 elemental areas. Assuming that one changeover from black to white per second corresponds to a frequency of 1 Hz and that a DIN-A4 page is transmitted e.g. in $6\frac{1}{2}$ min, the picture frequency is

$$\frac{\text{Number of elements}}{\text{Time} \times 2} = \frac{10^6}{6.5 \times 60 \times 2} \approx 1\,200\text{ Hz}$$

For the transmission over a telephone channel this signal is modulated with a carrier frequency of 1900 Hz. The bandwidth required by the facsimile signal is therefore 1900 $\pm$ 1200 Hz.

The maximum possible picture frequency (in theory) of a HELLFAX met chart recorder can be calculated from drum speed, drum diameter and index of co-operation:

$$f_{P\max} = M \times \pi \times \frac{n}{120}$$

where $f_{P\max}$ is the maximum picture frequency

n is the drum speed

M is the index of co-operation.

By way of example Table 6a gives some characteristics of a commercial-type HELLFAX met chart recorder, together with the picture frequencies.

Table 6a   Characteristics of a HELLFAX met chart recorder

n/rpm	M	$f_{P\max}$/Hz	l/lines mm^{-1}	Type of modulation
60		450	4.3	
90	288	675	or	Amplitude modulation
120		900	4.8	
60		900	3.8	
90	576	1300	or	Amplitude modulation
120		1800	8.6	

n	Drum speed
M	Index of co-operation
$f_{P\max}$	Maximum picture frequency
l	number of lines mm^{-1}

Table 6b   Characteristics of a HELL phototelegraph apparatus

n/rpm	M	$f_{P\,max}$/Hz	l/lines mm^{-1}	Type of modulation
60		415		
90	264	620	3.8	Frequency modulation
120		825		
60		550		
90	352	825	5.0	Frequency modulation
120		1100		

n	Drum speed
M	Index of co-operation
$f_{P\,max}$	Maximum picture frequency
l	Number of lines mm^{-1}

Transmission path

The received picture should be a copy-true reproduction of the transmitted original. This requirement imposes certain demands on the transmission characteristics of the radio links. The signals of a facsimile transmission may be distorted as the result of rapid delay variations of the transmission circuit (Fig. 30). In most cases, however, a facsimile transmission will contain sufficient redundancy to permit satisfactory interpretation of the received information.

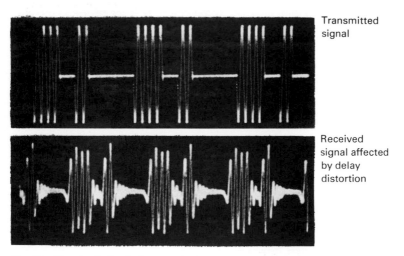

Transmitted signal

Received signal affected by delay distortion

Fig. 30   Signals of a facsimile transmission

52

As is known, nearly all transmission circuits are characterized by a certain bandwidth, i.e. an upper cut-off frequency which serves as the basis for determining the maximum possible picture frequency. The maximum picture frequency and the maximum signalling rate must be adapted to the transmission path. A certain measure of adaptability to different circuit capacities is an inherent feature of most facsimile units. Such units afford various settings as regards drum speed and index of co-operation. This will be necessary if the total transmission path is made up of a number of modulation sections, e.g. physical circuit and radio circuit. The physical circuit will then depend, among other things, on the bandwidth of the voice channel, which may be 3 kHz, and the radio circuit on the bandwidth of the radio transmitter and receiver. In the longwave and shortwave ranges the available bandwidth is determined by the frequency allocation scheme, i.e. the frequency spacing between the local transmitter and any neighbouring-frequency transmitters. Where physical and radio circuits are connected in tandem, the physical circuit should have the same bandwidth as the radio circuit.

In the longwave range the bandwidth may be limited by the antenna characteristics. In the shortwave range, in particular, interference phenomena must be taken into consideration which may compel the receiving station to select a smaller bandwidth. If the thickness of the traces and the size of the symbols are carefully selected it may be possible, in the case of weather charts for instance, to achieve satisfactory transmission quality in spite of a reduction of the bandwidth and the picture frequency.

The bandwidths required for facsimile transmissions, employing class of transmission F4, are compiled in Table 7 with the bandwidth being defined

Table 7 Picture frequency and bandwidth of a transmission channel

Picture frequency $f_{P\,max}/$Hz	Bandwidth $B/$Hz
450	$\geqslant 900$
675	$\geqslant 1350$
900	$\geqslant 1800$

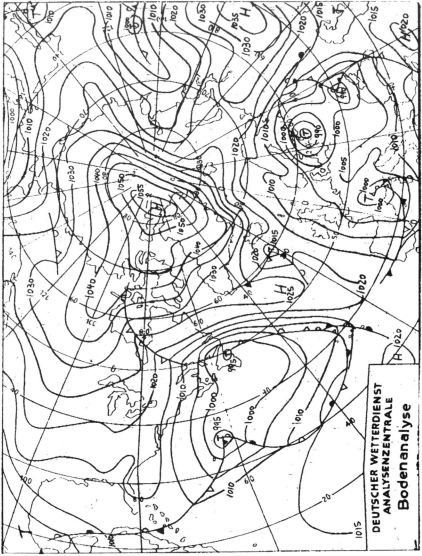

Fig. 31   Weather chart transmitted over a radio link

approximately as

$$B \geqslant 2f_{P\,max}.$$

For the transmission of black and white copies, the interposed modulators of the radio transmitters and the demodulators of the radio receivers must also be taken into consideration. They must be suitable for the picture frequency to be transmitted. Figure 31 shows a weather chart transmitted over a radio path.

Facsimile information is today transmitted by frequency keying of the HF carrier (F4) with a frequency shift of ± 150 Hz in the longwave range and ± 400 Hz in the shortwave range.

With phototelegraphy – the transmission of pictures, photographs, etc. – the halftones (gray tones) are transmitted in addition to black and white values. Linear modulation and demodulation methods must be adopted for this purpose. The black and white values determine the maximum bandwidth. The modulation methods are amplitude as well as frequency modulation with a variable frequency shift of up to ± 400 Hz. Table 6 shows some characteristics of a HELL phototelegraph equipment. The single-sideband method with suppressed carrier (A3J) is employed for picture transmission over shortwave radio links.

2.6 Types of Traffic

Telegraph and data communication over radio links is characterized by various types of traffic between the subscribers. The distinction is made between dedicated circuits – also called point-to-point circuits – and telex or gentex circuits.

The exchange of information is possible in various ways. The types of traffic are represented in Fig. 32.

▶ If messages are transmitted in one direction only, we speak of unidirectional or **one-way traffic.** Between the subscribers there is no need for a direct reply. Broadcast operation also comes under this heading.

▶ If, on the other hand, signals are transmitted on a two-way link, we speak of **half-duplex traffic.**

▶ Over many radio links, simultaneous exchange of information in both directions is required. For this purpose, two-way links must be provided. This mode is known as **duplex traffic.** Messages are transmitted in each direction independent of each other.

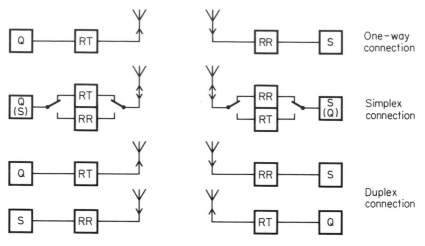

Fig. 32 Types of traffic

RT   Radio transmitter   Q   Source
RR   Radio receiver   S   Sink

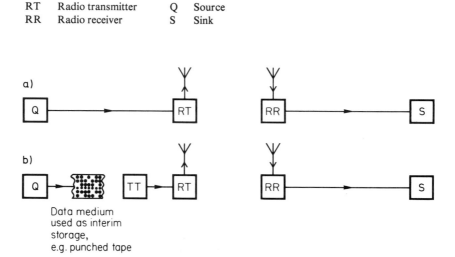

Fig. 33 Mode of operation for direct and indirect communication
(a) On-line operation (direct connection)
(b) Off-line operation (indirect connection)

TT   Tape transmitter
RT   Radio transmitter
RR   Radio receiver
Q   Source
S   Sink

Regarding the mode of operation, the distinction is made in radio-telegraph and data transmission between **on-line** and **off-line operation** (Fig. 33).

In on-line operation, the messages are passed without buffering direct to the radio transmitter. The information source is connected to the transmitter.

In off-line operation the incoming information is first buffered – in punched tape, for example – before being forwarded to the transmitter.

2.7 Message Multiplexing

In radio-telegraph communication there is often a need to simultaneously transmit several telegraph messages and data over **one** radio link. This is made possible by multiple utilization of the voice-band between 0.3 and 3 kHz as provided by radio transmitters and receivers.

With the aid of voice-frequency telegraph systems (see Chapter 5) these voice-grade channels can be split up into a number of narrow-band channels (**frequency-division multiplex** method, Fig. 34(b)).

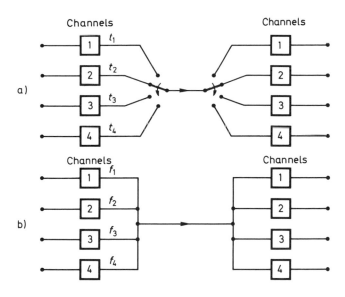

Fig. 34   (a) Time-division multiplex method (principle)
(b) Frequency-division multiplex method (principle)

The advantage of this method is that the messages on the individual channels are independent of each other; they can be transmitted at different bit rates depending on the channel bandwidth.

Information from several different channels can also be sequentially interleaved during transmission (**time-division multiplex method,** Fig. 34(a)). In order to retain the bit rate of the channel it is necessary to shorten the pulses on the transmission path. For example, when two channels each with a speed of 50 bauds are combined, the information flows over the communication circuit at 100 bauds. For this purpose, the individual pulses must be fitted into a common timing pattern.

Frequency and time division multiplex methods are basically equivalent in their utilization of a frequency band.

Similarly to the frequency and time division multiplex methods the individual bits of a code combination can also be transmitted in parallel over several channels with different frequencies or in series one after the

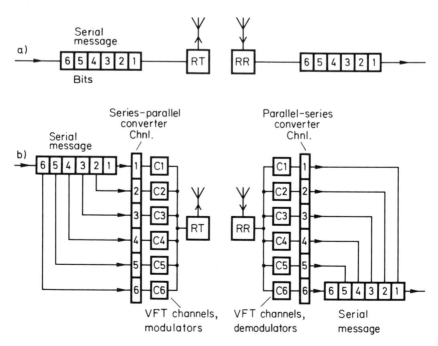

Fig. 35   (a) Series and (b) parallel transmission of telegraph signals

58

other over a single channel. Over shortwave radio links, the signals are usually transmitted serially.

The transmission of binary-coded information over shortwave links is, however, limited to speeds not exceeding 200 bauds because of the inherent interference, particularly fluctuation of the propagation delay due to a change in the reflecting layers. The delay time variations may amount to about 2 ms, sometimes even 3 ms. They therefore exceed half the pulse length of a 5 ms telegraph code element at 200 bauds.

A telegraph pulse is normally sampled in the centre of its duration. With the named delay fluctuation, distortion increases to such an extent that the code element is no longer correctly interpreted. This compels us to set the maximum transmission speed for serial transmission over shortwave radio links in such a way that the pulse duration is limited to 5 ms.

Parallel transmission of the individual bits of a code combination is possible for higher transmission speeds while retaining the bit length of, for example, 10 ms [11]; this is effected by transmitting the bits simultaneously in several channels (Fig. 35).

3 Types of Modulation

For the long-distance transmission of telegraph data signals, only alternating current is used as a carrier.

In order that signals may be transmitted over a radio path, they are impressed with the aid of a modulator on an a.c. carrier wave in the HF position. In telegraph transmitters, this process is known as modulation or **keying.** In modulating the carrier, either the amplitude, the frequency or the phase is changed in accordance with the binary signal states.

At the receiving end, the original signal must be recovered, i.e. demodulated. The interference to which the transmission path is exposed manifests itself in changes in the received signal. The adverse transmission conditions on radio links make it necessary to use special facilities (see Chapter 4).

For telegraph and data transmission, suitable types of modulation, also known as types of transmission, have been evolved, in order to permit maximum transmission reliability of the radio path. A type of transmission is the method that is used for modulating the telegraph signals onto the high-frequency carrier. The designations for the types of transmission are standardized by international convention and are laid down in accordance with the Radio Regulations, VO Funk Art. 2, Section I, Geneva 1968. The designations for frequency-modulation (FM) and amplitude-modulation (AM) transmission are listed in Tables 8 and 9. In shortwave telegraph and data transmission the types of transmission A1, A2, A3A, A3J, A3B, A7A, A7B, A7J and A9B are usual for amplitude modulation. With frequency modulation, the types of transmission F1, F4 and F6 (frequency-shift keying FSK) are of significance. Figure 36 illustrates the types of transmission most frequently used in radio telegraphy.

There now follows a brief survey of the various types of transmission used in radio telegraphy, together with their advantages and disadvantages.

3.1 Amplitude Modulation (AM)

In type of transmission A1 – also called CW telegraphy – the high-frequency carrier is keyed on and off in the cadence of the telegraph signals.

Table 8   Designations of the types of transmission (extract from Radio Regulations VO-Funk-Art. 2, Section I)

Types of modulation of main carrier	Symbol
(a) Amplitude	A
(b) Frequency (or Phase)	F
(c) Pulse	P

Types of transmission	Symbol
(a) Absence of any modulation intended to carry information	0
(b) Telegraphy without the use of a modulating audio frequency	1
(c) Telegraphy by the on-off keying of a modulating audio frequency or audio frequencies, or by the on-off keying of the modulated emission (special case: an unkeyed modulated emission)	2
(d) Telephony (including sound broadcasting)	3
(e) Facsimile (with modulation of main carrier either directly or by a frequency modulated sub-carrier)	4
(f) Television (vision only)	5
(g) Four-frequency diplex telegraphy	6
(h) Multichannel voice-frequency telegraphy	7
(i) Cases not covered by the above	9

Supplementary characteristics	
(a) Double sideband	(none)
(b) Single sideband:	
▶ reduced carrier	A
▶ full carrier	H
▶ suppressed carrier	J
(c) Two independent sidebands	B
(d) Vestigial sideband	C
(e) Pulse:	
▶ amplitude modulated	D
▶ width (or duration) modulated	E
▶ phase (or position) modulated	F
▶ code modulated	G

Notes

The transmissions are characterized by the type of modulation of the main carrier, type of transmission and supplementary characteristics.

The width of a radio channel is given by stating the bandwidth in kHz before the designation of the type of modulation.

e.g.   6A3J (single-sideband 6 kHz)
12A3B (two independent sidebands each of 6 kHz).

Table 9   The usual types of transmission (extract from VO-Funk-Art. 2, Section I)

Type of Modulation of Main Carrier	Type of Transmission	Supplementary Characteristics	Symbol
Amplitude Modulation	With no modulation	—	A0
	Telegraphy without the use of a modulating audio frequency (by on-off keying)	—	A1
	Telegraphy by the on-off keying of an amplitude-modulating audio frequency or audio frequencies, or by the on-off keying of the modulated emission (special case: an unkeyed emission amplitude modulated)	—	A2
	Telephony	Double sideband	A3
		Single sideband, reduced carrier	A3A
		Single sideband, suppressed carrier	A3J
		Two independent sidebands	A3B
	Facsimile (with modulation of main carrier either directly or by a frequency modulated sub-carrier)	—	A4
		Single sideband, reduced carrier	A4A
	Television	Vestigial sideband	A5C
	Multichannel voice-frequency telegraphy	Single sideband, reduced carrier	A7A
	Cases not covered by the above, e.g. a combination of telephony and telegraphy	Two independent sidebands	A9B
Frequency (or Phase) Modulation	Telegraphy by frequency shift keying without the use of a modulating audio frequency: one of two frequencies being emitted at any instant	—	F1

Table 9—*Continued*

Type of Modulation of Main Carrier	Type of Transmission	Supplementary Characteristics	Symbol
Frequency (or Phase) Modulation	Telegraphy by the on-off keying of a frequency modulating audio frequency or by the on-off keying of a frequency modulated emission (special case: an unkeyed emission, frequency modulated)	—	F2
	Telephony	—	F3
	Facsimile by direct frequency modulation of the carrier	—	F4
	Television	—	F5
	Four-frequency diplex telegraphy	—	F6
	Cases not covered by the above, in which the main carrier is frequency modulated	—	F9

This type of transmission is still sometimes used for Morse operation. The generated frequency spectrum is distributed symmetrically about the carrier frequency; it depends essentially on the telegraph speed and the shape of the signals [3]. Due to the simplicity of this modulation method, type of transmission A1 was for a long time the most commonly used telegraph and radio transmission method. Modulation in the radio transmitter is very simple, and little equipment is needed for demodulation at the receiving side. The drawback of amplitude modulation in general is that the characteristic instants of the signal are level-dependent and therefore prone to distortion with amplitude fluctuation. Especially in the inter-signal intervals, impulse-type noise may simulate telegraph signals.

In the A1 type of transmission the high-frequency carrier is keyed in line with the telegraph signals, i.e. at the start of each signal the HF amplitude rises to its maximum and drops to zero again at the end of the signal (Fig. 21). With square-wave signals the **keying is hard,** which results in a broad keying spectrum that in certain cases may be several kHz wide, thus causing considerable interference with adjacent radio services. If the

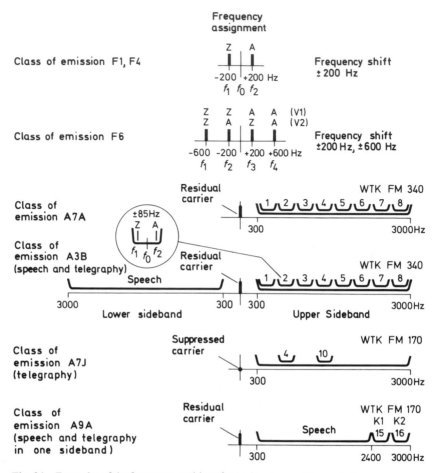

Fig. 36   Examples of the frequency positions for various types of transmission

telegraph pulses rise slowly, we speak of **soft keying.** The rise and fall times of the telegraph signals usually each amount to 20% of the total character length. Figure 37 shows a telegraph signal with A1 soft keying.

In the A2 type of transmission the carrier is modulated with a voice frequency of 500 or 1000 Hz, for example, and the modulation is switched on and off in accordance with the characteristic states of the telegraph signals. (This type of transmission is fast declining in importance.)

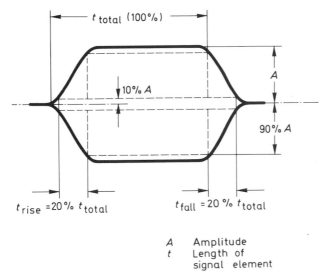

A Amplitude
t Length of
signal element

Fig. 37   Telegraph signals with A1 soft keying

A Amplitude
t Signal length

In the A3A, A3B and A3J types of transmission it is possible, through the use of **single-sideband amplitude modulation,** to make available one or more voice-grade channels for radio-telephony. The frequency range of one such channel lies between 300 and 3000 Hz according to CCIR. With the frequency-division multiplex method it is possible to load these voice-grade channels with groups of narrow-band channels for voice-frequency telegraphy (WTK) at speeds up to 200 bauds (Chapter 5). When the sidebands are occupied by channels of a voice-frequency telegraph system, the A3A, A3B and A3J types of transmission are given the designations A7A, A7B and A7J. In types A3J and A7J the carrier is suppressed ($\geqslant 40$ dB acc. to CCIR), so that the total power of the transmitter is available for the information content of the sideband.

3.2 Frequency Modulation (FM)

For data transmission at up to 200 bits/s frequency modulation – frequency-shift keying – is generally used today [2]. Frequency modulation is less affected by amplitude fluctuation of the received signal; it has the further advantage of being less prone to interference, since each of the shift frequencies is transmitted with the full power of the radio transmitter. This is why this modulation method was adopted early in the history of shortwave radio transmission.

Where just one or two information channels are to be operated, the types of transmission which are mostly used are F1, F4 and F6.

In type F1, two adjacent frequencies $f1$ and $f2$ are assigned to the characteristic states Z and A (stop- and start-polarity) of the binary signal, and the high frequency of the radio transmitter is shifted between these two states (shift frequencies). The mid-frequency is expressed as

$$f_0 = \frac{1}{2}(f_A + f_Z)$$

and the frequency shift (Fig. 38) as

$$H = \frac{1}{2}(f_A - f_Z)$$

In accordance with CCITT and CCIR Recommendations, the frequency f_Z is always the lower of the two. In shortwave and longwave radio communication, frequency shifts of ± 50 Hz to ± 200 Hz are usual. In older systems, frequency shifts of ± 400 Hz and more are also still in use.

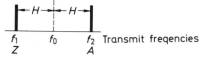

a) Frequency assignment for F1 and F4

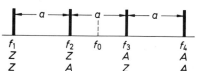

b) Frequency assignment for F6

Fig. 38   Frequency allocation with types of transmission F1, F4 and F6

f Frequency
H Shift
a Shift frequency spacing
A Start polarity
Z Stop polarity

66

Facsimile equipment is operated with type of transmission F4. As in the case of F1, the high frequency of the radio transmitter is keyed between the two shift frequencies in accordance with the black—white scanning signal from the original document. Due to the large number of picture points in facsimile transmission, it is usual over shortwave links to use a frequency shift of ±400 Hz [5, 6] and a transmission speed up to 3600 bauds. For picture transmission the gray tones are represented by frequency shifts of different magnitudes.

With F6 transmission, two completely independent messages can be transmitted simultaneously over a radio link by shifting between four adjacent frequencies. The frequencies from the radio transmitter are allocated to the characteristic states of the binary signals of the two channels V1 and V2 as shown in Fig. 38. The CCIR recommends a frequency spacing a between the shift frequencies of 400 or 200 Hz in the shortwave range and 100 or 50 Hz in the longwave range [8]. It is best to distribute the individual information pulses of the channels V1 and V2 into equal-timing patterns with the aid of a pulse allocator, as otherwise information pulses of the independent signals may occur with any degree of brevity which, since the bandwidth of the transmission channel is prescribed, may lead to distortion. In order to avoid this, single-sideband modulation with voice-frequency multiplexing (Chapter 5) is increasingly replacing F6 transmission.

3.3 Types of Modulation for Data Transmission at 2400 bits/s

In order to permit data transmission at a rate of 2400 bits/s on a voice-grade channel over a shortwave radio link, special matching of the transmission method to the conditions prevailing in the transmission medium is necessary. We know that multipath propagation in shortwave transmission causes delay fluctuation in the order of several milli-seconds, so that a minimum pulse duration of 5 ms is necessary for trans-mission (see also Section 2.7). Because of the considerably shorter bit length, serial data transmission at 2400 bits/s over an SW voice-grade channel is in this case not possible. This statement is not contradicted by the fact that facsimile transmission is practised over shortwave links, since the redundancy of the image permits legible reproduction even with high-delay distortion (see Section 2.5). A solution that remains for the transmission of coded information is to split the high-speed serial channel into a number of low-speed parallel channels.

In the past few years, a number of methods have been developed which permit the transmission of high bit rates over SW radio links by dividing them among several channels within the speech channel. The last CCIR Plenary Assembly (1974), for example, recommended a system [11] by which data can be transmitted with a high bit rate over several parallel FM channels. In this system, the data arriving serially are divided by a series-parallel converter among 12 parallel channels. Transmission is effected by means of the shortwave voice-frequency telegraph channels (WTK) set down some time ago in a CCIR Recommendation [9] (see Chapter 5), each of which permits synchronous transmission at a rate of 100 bauds, taking as a basis the 170 Hz channel spacing scheme. The disadvantage of these methods, however, is that they only permit a transmission speed of 1200 bits/s in a voice-grade channel at relatively great expense.

As regards optimum utilization of the voice-frequency band, considerably better results are obtained with an SW data transmission system developed ten years ago, which employs 20 voice frequencies with an equal spacing of 110 Hz and an additional synchronizing tone. By using four phase positions, each of the 20 voice frequencies can be loaded with two mutually independent bits (dibits) each with a modulation rate of 75 bauds, so that the maximum possible bit rate is $2 \times 20 \times 75$ bits/s $= 3000$ bits/s.

For shortwave links with a transmission speed of 2400 bits/s the four-phase **coherent phase-shift** keying method has proved to be most advantageous. The advantage of this type of modulation over the widespread two-frequency shift keying method used today for shortwave communication is an improvement in the signal-to-noise ratio under white noise conditions by about 1 dB.

When choosing the modulation method, use was made of a fact that has a bearing on the choice of channel spacings in frequency diversity operation: namely, a number of investigations have shown that two closely spaced WTK channels will both be affected simultaneously by multipath propagation, yet more widely spaced channels (> 400 Hz) are affected independently of each other. For this reason, a channel spacing of at least 400 Hz is required for frequency diversity operation. On the other hand, since two adjacent transmission channels differ only slightly, both in phase distortion and in their significant intervals, the most suitable SW data transmission method is a phase shift keying method that makes use of the 'stability' of the phase difference between closely spaced voice frequencies. This modulation method is referred to as **frequency-differential phase-shift keying,** in contrast to time-differential phase-shift keying, wherein the

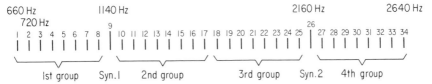

Fig. 39 Arrangement of the transmission frequencies within the speech band of a 2400 bits/s transmission system

information content lies in the phase difference between two consecutive significant intervals. With frequency-differential phase-shift keying the information exists in the phase difference between two adjacent voice frequencies.

In order to permit data transmission with a modulation rate of 2400 bits/s in the speech channel of a shortwave radio link we require – as mentioned above – several modulated frequencies with which the bit stream, divided by a series–parallel converter, is transmitted at correspondingly lower speeds. The modulation rate can be increased even further by using not binary but quaternary phase-shift keying. By modulating the information with four phase positions of a voice frequency, each displaced by 90° from the next, the interference immunity of the transmission channel compared with that of binary phase-shift keying (0/180°) can be retained by doubling the transmitting power; this is reasonable considering that the information flow is also doubled.

Through appropriate choice of modulation interval length and spacing of the voice frequencies, the delay fluctuation caused by multipath propagation and the sources of error connected with it can be avoided to a great extent. If, at the same time, we use four-frequency differential phase shift keying and a total of 24 parallel voice-frequencies, each of which is phase-modulated at a rate of 50 bauds, we obtain the desired modulation rate of $2 \times 24 \times 50$ bits/s $= 2400$ bits/s.

Two voice frequencies are provided in addition for the synchronization of the demodulator. The forward error correcting method may be employed to reduce the error rate. For this purpose another eight frequencies are required. Figure 39 shows the assignment of these channels within the voice band for a 2400 bits/s transmission terminal, adjacent frequencies being spaced at 60 Hz.

69

4 Structure of a Shortwave Radio Link

Figure 40 shows the schematic diagram for the transmission of digital signals over a radio link. The terminals may be teleprinters, data terminal equipment or facsimile apparatus, serving in each case as either transmitting equipment (source) or receiving equipment (sink). The modulator (Mod) impresses the digital signal on a carrier wave. At the receiving end, the demodulator (Dem) converts the carrier-frequency signal back into digital form.

When data protection methods are applied, coding facilities (Cod) and decoding facilities (Dec) are provided at the transmitting and receiving ends, respectively. Data protection methods are described in Chapter 7.

Diversity procedures are used in order to mitigate the adverse effects of severe fluctuation in the transmission characteristics of the radio link. The signal reaches the receive side by different paths, and by selection or combination of the transmitted signals the received message will be to a large extent error-free. Multiple transmission of the message at the send side is also possible. The diversity methods are marked Div in the diagram and are discussed in Chapter 6.

The radio transmitter RT and the radio receiver RR with their associated antennae and the interstitial medium form the radio transmission circuit.

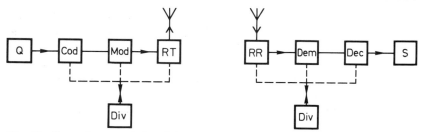

Fig. 40   Set-up for transmission of digital signals over a radio link

Cod, Dec	Data-protection equipment with error correction
Div	Equipment for implementation of diversity procedures
Q	Source
S	Sink
RT	Radio transmitter with antenna
RR	Radio receiver with antenna
Mod, Dem	Modulator and demodulator for the digital signal

4.1 The Radio Equipment

The quality of the entire radio link is dependent on the radio equipment. The transmitter and receiver and their antennae, the transmitting power and the frequency stability are all factors of crucial importance for the quality of the radio link. The quality is also influenced by the propagation conditions on the shortwave radio link, which in turn depend on the distance and the frequencies used.

In the following section some important characteristics of radio equipment are named, which concern telegraphy and data transmission.

The telegraph and data interfaces between the binary modulator and the radio transmitter and between the radio receiver and the binary demodulator must be matched as regards impedance and signal level. The IF interface is mostly used for frequency-shift modulation and demodulation (types of transmission F1, F4, F6). The transmit converter in the radio transmitter, and the radio receiver should be furnished with bandpass and low-pass filters matched to the transmission speed and having the minimum possible attenuation and delay distortion.

For single-sideband modulation (types of transmission A7A, A7B or A7J) the VF interface is used for connection to the telegraph transmission terminals. Here, the harmonic response of the VF channel of the radio equipment is also of significance; a harmonic distortion factor of better than 1% at 1000 Hz is desirable. When one sideband is modulated with several tones, e.g. in SW VFT systems (WTK), difference and summation tones result in intermodulation distortion, which has an adverse effect on the transmission characteristics of the individual narrowband channels. The intermodulation distortion attenuation within the HF and IF bands should be greater than 50 dB with two signals. If an AGC amplifier is connected in series with the line on the transmitting side for type of transmission A3, so as to keep the modulation of the transmitter as even as possible, additional non-linear modulation products may be produced due to overdriving in WTK operation.

When WTK channels are employed in diversity operation or multichannel data transmission systems are operated over single-sideband links, the group delay distortion of the overall transmission channel must be taken into account. This group delay distortion is due to the fact that not all of the frequencies of the frequency band used for shortwave transmission are transmitted at the same velocity or in bit-synchronous fashion between the transmitting and the receiving station.

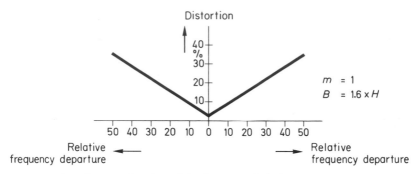

Fig. 41 Distortion as a function of the frequency deviation of a radio channel referred to the frequency shift

$$Distortion = k \left(\frac{v}{B}\right)\left(\frac{\Delta f \times 100}{H}\right)$$

v Telegraph speed
B Bandwidth
H Frequency shift
m Modulation index $= \dfrac{\text{Frequency shift}}{\text{Telegraph speed}}$
f Frequency error
k Constant

For frequency-shift keying, the stability of the carrier frequencies in the radio equipment is particularly important. Figure 41 shows the distortion occurring in a radio channel as a function of the frequency deviation referred to the frequency shift. The distortion is almost proportional to the frequency error Δf and the telegraph speed v and inversely proportional to the bandwidth B and the frequency shift H.

High-frequency accuracy of the radio channel [10] permits low-distortion transmission of telegraph and data signals. Frequency deviation by 10% of the frequency shift used in type of transmission F1 results in a 7% increase in distortion, as shown by the curve in Fig. 41. Through the use of crystals or frequency synthesizers in the radio equipment, high-frequency stability is obtained. The deviation from a set frequency in this radio equipment can be reduced to something of the order of 3×10^{-8}. The frequency accuracy over a radio link can therefore be better than 2 Hz over the whole range up to 30 MHz. With types of transmission F1, F4 and F6, high-frequency stability of the radio equipment permits the use of small frequency shifts and hence bandwidths matched to the telegraph speed. In this way, the

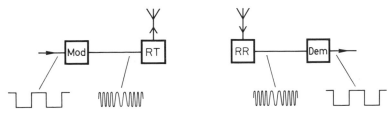

Fig. 42   Modulator and demodulator of the radio equipment

signal-to-noise ratio of the transmission channel is increased and the bit error rate reduced.

The frequency stability here necessary for radio equipment is laid down in a CCIR Recommendation [10].

For any given transmitting or receiving frequency the **setting accuracy** of the radio equipment determines the maximum frequency deviation with which this frequency can be selected on the radio equipment.

4.2 Modulation and Demodulation Equipment

For the transmission of telegraph messages and data, the binary signals are normally converted by means of a modulator (Mod, Fig. 42) into voice-frequency signals and then passed to the radio transmitter. On the receive side, the radio receiver feeds the voice-frequency signals to the demodulator (Dem), which converts them back into the original direct-current signals. Figure 43 shows a radio receiver installation consisting of radio receiver, telegraph demodulator and teleprinter. The methods preferred today for telegraph and data transmission over shortwave radio links are, because they are less susceptible to interference, frequency-shift keying or single-sideband modulation with special voice-frequency telegraph systems (WTK). The great variation of the received signal due to fading demands a maximum of selectivity together with a wide AGC range of the demodulator. The frequency shifts, telegraph speed and filter bandwidth have to be matched to each other in order to ensure optimum transmission.

In order to remain independent of the transmission frequency in the longwave or shortwave range, as the case may be, during modulation in the transmitter and demodulation in the receiver, the output of the modulator

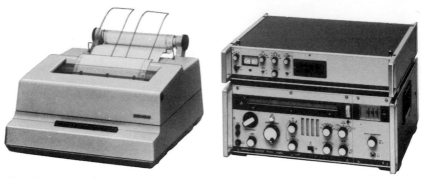

Fig. 43   Radio receiver equipment for F 1, F4 and F6 transmissions

and the input of the demodulator for frequency-shift keying usually operate with a single standard transfer frequency, usually 30 kHz, for interfacing with the radio equipment.

4.3 Modulation Equipment of a Radio Transmitter

Figure 44 is a schematic representation of the modulation equipment of a radio transmitter for preparing telegraph and data signals for the A1, A2, F1, F4 and F6 types of transmission. Direct-current or voice-frequency signals can be applied to the input via the internationally standardized telegraph and data interfaces. The modulated 30 kHz wave is fed to the frequency converter of the radio transmitter for further processing. In F6 transmission the two signals of channels V1 and V2 are decoupled by inter-posed d.c./VF converters CON. The VF keying signals are demodulated by the tone demodulator TDM. For type of transmission F6 the d.c. signals of channels V1 and V2 from amplifier stages V are combined in the F6 adder F6–A. In order to limit the frequency spectrum produced by keying, the signals pass through the low-pass keying filter L to the modulators A1 Mod, A2 Mod, FMO. In types of transmission F1, F4 and F6 the signals key the 30 kHz wave into the channel frequency. In the A1 and A2 types of transmission, a 30 kHz oscillation supplied by the frequency converter is keyed on and off in the cadence of the telegraph signals by the A1 or A2 modulator.

Both for telephony and for telegraph and data transmission, the information signals are changed by the frequency converter of the radio transmitter from the 30 kHz position to the transmission frequency position. In modern

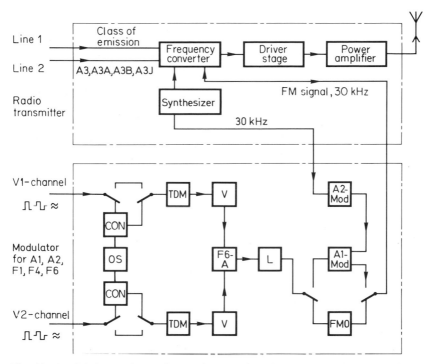

Fig. 44   Modulation equipment in a radio transmitter for types of transmission A1, A2, F1, F4, F6

OS	Oscillator	L	Low-pass keying filter
CON	d.c./VF converter	A2-Mod	A2-modulator
TDM	Tone demodulator	A1-Mod	A1-modulator
V	Amplifier	FMO	FM modulator
F6-A	F6 adder		

transmitters, the necessary crystal-stabilized frequency is supplied by a frequency synthesizer. The generated HF signal then passes via the driver stage and power amplifier to the antenna.

4.4 Demodulation Equipment at the Receive Side

The operating principle of demodulation equipment for types of transmission F1, F4 and F6 is shown in the block diagram (Fig. 45). Modern

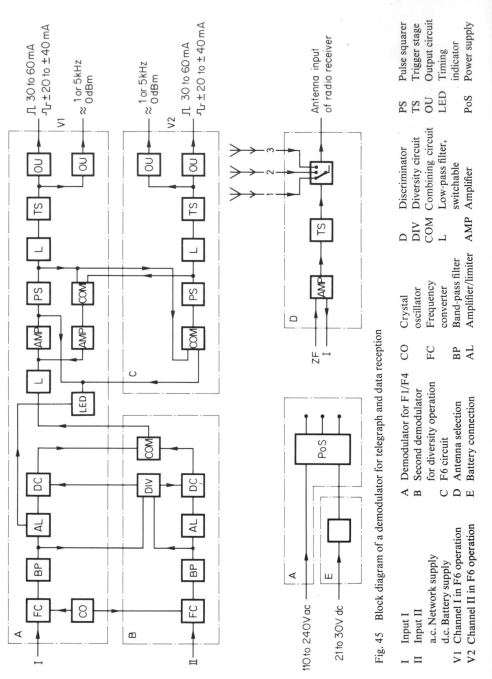

Fig. 45   Block diagram of a demodulator for telegraph and data reception

I   Input I
II   Input II
a.c. Network supply
d.c. Battery supply
V1 Channel I in F6 operation
V2 Channel II in F6 operation

A   Demodulator for F1/F4
B   Second demodulator
    for diversity operation
C   F6 circuit
D   Antenna selection
E   Battery connection

CO   Crystal
     oscillator
FC   Frequency
     converter
BP   Band-pass filter
AL   Amplifier/limiter

D   Discriminator
DIV Diversity circuit
COM Combining circuit
L   Low-pass filter,
     switchable
AMP Amplifier

PS   Pulse squarer
TS   Trigger stage
OU   Output circuit
LED Timing
     indicator
PoS Power supply

Fig. 46   Telegraph demodulator FSE 401 A with built-in antenna selection unit

telegraph demodulator is shown in Fig. 46. The equipment permits demodulation of FM telegraph, data and facsimile signals with frequency shifts of ± 20 Hz to ± 1500 Hz offered by a radio receiver in the IF position at 30–1400 kHz or in the VF position at 1.9 kHz. At its output the equipment supplies direct-current or voice-frequency signals over internationally standardized telegraph and data interfaces for the connection of teleprinters, facsimile and crypto equipment or data protection systems.

The FM signals applied to the input are converted into an intermediate frequency of 10 kHz in the frequency converter FC with auxiliary oscillator CO. The frequencies of the received FM channel are next selected by a band-pass filter BP that is variable from ± 75 Hz to ± 1500 Hz. A multistage amplifier AL then limits the amplitude (limitation range about 50 dB) of the signals. The frequency-modulated signals are now demodulated in a linear discriminator DC. The discriminator is followed by a switchable low-pass filter L, whose cut-off frequency can be selected in six steps for telegraph speeds from 50 to 3600 bauds. Via a differential amplifier AMP the signals reach a pulse squarer PS where they are amplified and limited. The following flip-flop TS steepens the edges of the d.c. signals even further. These signals then drive the output circuits OU. The interface to the terminal equipment may be designed for high-level signals (Fig. 45) or for low-level signals as per V.28 [18].

The channels V1 and V2 in F6 operation are separated after demodulation. The discriminator DC converts the four possible frequency conditions into corresponding 'voltage steps'. Since channel V1 in F6 transmission corresponds to the channel of F1 transmission, the signals for channel V1 are acquired as in F1 operation. The information for channel V2 is formed by taking the difference between the V1 channel COM and the linear demodulated signal.

77

As described for F1 operation, the signals pass through low-pass filters and pulse squarers to the output circuits of the channels V1 and V2.

For space diversity operation (see Chapter 6) with a second radio receiver, a second complete demodulation path with diversity circuit is necessary. It can be incorporated in the same equipment. For this type of operation, the frequency converters FC of both demodulation paths are fed by a common oscillator CO. The signals arriving over the two receive paths are evaluated (DIV) before limitation. After demodulation, a process of maximal ratio post-detection diversity combination and selection is applied. If the difference between the receive levels is only slight, the signals from the two paths are added (COM); if the levels differ by more than 8 dB, the receive path with the weaker signal is cut off.

Instead of the space diversity method using two radio receivers, the antenna selection method can be employed, whereby two or three antennae and only

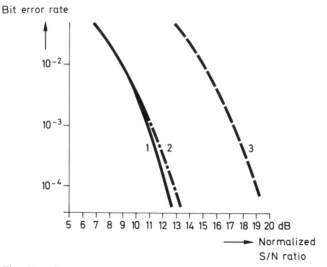

Fig. 47   Bit error rate as a function of the standard signal-to-noise ratio with frequency modulation for the demodulator as per Fig. 45

Curve 1:   Ideal receiver (as per CCIR Report 195)
Curve 2:   F1 operation, shift ± 40 Hz, bandpass filter 10 kHz ± 60 Hz,
   low-pass filter 100 bauds (cut-off frequency f_c = 80 Hz)
   telegraph speed 100 bauds
Curve 3:   F1 operation, shift ± 400 Hz, bandpass filter 10 kHz ± 700 Hz,
   low-pass filter 200 bauds (cut-off frequency f_c = 160 Hz)
   telegraph speed 200 bauds

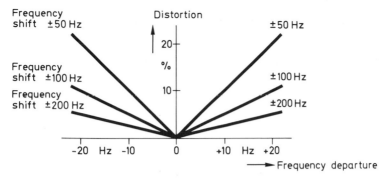

Fig. 48   Distortion as a function of frequency deviation in F1 operation with 50 bauds and 1 : 1 reversals

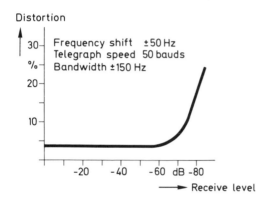

Fig. 49   Distortion as a function of input level in F1 operation

one radio receiver are operated. The selection circuit automatically connects the radio receiver to the antenna located where field strength is best. If the IF voltage of the radio receiver falls below a set threshold value, selection is initiated via a trigger stage.

Frequency deviation of up to ±500 Hz on the radio path can be automatically compensated for by a frequency error corrector in the system shown in Fig. 45. Figures 47, 48 and 49 show the characteristics of this demodulation equipment in F1 operation, namely the bit error rate as a function of signal-to-noise ratio, the signal distortion as a function of the possible frequency deviation and the signal distortion as a function of the input level. An LED tuning indicator permits the radio receiver to be tuned to a given frequency with the required frequency swing.

79

5 Transmission Equipment for Single-sideband Radio Links

Shortwave radio equipment is capable, through the use of single-sideband modulation in types of transmission A3A, A3B and A3J, of providing voice-grade communication channels occupying the range from 300 to 3000 Hz. A voice-grade channel of this kind can be used for the simultaneous and independent transmission of telegraph messages and data over a number of narrow-band channels by means of a VFT system for shortwave links (WTK), for better utilization of the channel. Figure 47 shows the block diagram of a WTK setup. The front view of two typical WTK terminals is shown in Figs 48 and 49.

5.1 Voice-frequency Telegraph Terminal for Shortwave Radio Links (WTK)

Voice-frequency telegraph communication permits multiple utilization of a transmission circuit through frequency-division multiplexing. In departure

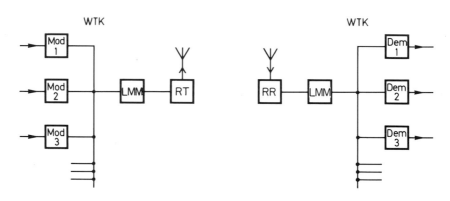

Fig. 50   Block diagram of a WTK transmission terminal

Mod	Modulator	RT	Radio transmitter
Dem	Demodulator	RR	Radio receiver
LMM	Line matching module		
	(for matching to the interface)		

Fig. 51 VFT transmission terminal WTK 1000 M for mobile applications

Fig. 52 VFT terminal WTK 1000 with eight send and eight receive channels

from the two-tone principle used earlier, the frequency-shift keying method used today is far less affected by level fluctuations. In FM transmission, each of the two signal states of the digital signals on the transmission path is assigned a voice frequency, so that by using different channel frequencies a number of mutually independent connections can be accommodated in one voice-grade channel. The individual WTK channels lie in the band between 300 Hz and 3000 Hz and thus correspond to the voice band of the radio channel.

The WTK frequency allocation, with channel spacings of 170 and 340 Hz, permits the accommodation of 16 and 8 channels, respectively, in the voice

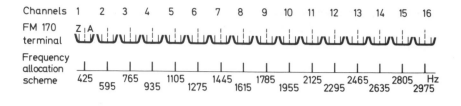

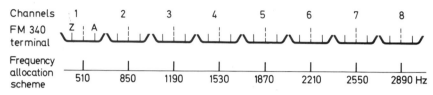

Fig. 53   Frequency allocation scheme of the WTK systems FM 170 and FM 340

1 to 16   Channels
A   Start-polarity current
Z   Stop-polarity current

band between 300 and 3000 Hz. Figure 53 shows the channel arrangement with the associated channel frequencies.

The carrier frequencies are odd-numbered multiples of 85 Hz. The frequency allocation scheme of the WTK system FM 170 follows a CCIR Recommendation [9, 13]. The frequency allocation and shift of the system FM 340 are based on those of the CCIR Recommendation for the system FM 170. A voice-grade channel can be loaded fully or partly with channels from one or both frequency allocation schemes. The most important WTK system data are listed in Table 10.

Over wire circuits, VFT systems are operated with channel mid-frequency spacings of 120, 240 and 480 Hz, the mid-frequencies being odd multiples

Table 10   Fundamental characteristics of WTK systems

| WTK-System | Channel spacing | Frequency shift | Maximum telegraph speed | |
| | | | Start/Stop | Synchronous |
	in Hz	in Hz	in bauds	in bauds
FM170	170	±42.5	75	100
FM340	340	±85	150	200

82

of 60 Hz. It should be pointed out that VFT systems designed for wire circuits are unsuitable for shortwave links. WTK systems with frequency modulation must satisfy more stringent conditions, the following in particular:

▶ Due to selective fading, considerable level differences (up to 40 dB) between the individual channels may occur. This is taken into account by the receive filters with a stop-band attenuation of at least 60 dB, so that in the case of selective fading the distortion in the individual channels caused by crosstalk between adjacent channels is kept to a minimum.

▶ The limitation in each demodulator must be rated for about 40 dB because of the great level fluctuations due to fading.

▶ To improve transmission quality, facilities must be provided whereby any two channels in different frequency positions can carry the same signal (frequency diversity operation). A signal transmitted on one channel can also be received on two radio receivers placed some distance apart (space diversity).

The modulation equipment of the WTK is therefore very similar in structure, apart from the different tuning components, to the modulators of the FM VFT equipment for transmission over physical circuits, yet the

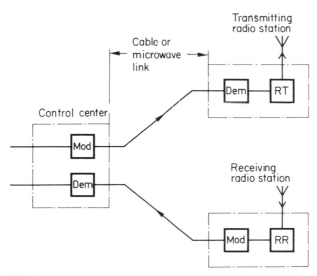

Fig. 54   WTK system used over feeder circuits between control centre and radio stations

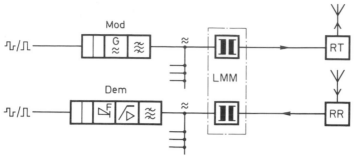

Fig. 55   Operating principle of a WTK channel
LMM   Line matching module

WTK receiving equipment, including diversity circuits, is more complex and expensive.

With its range of transmission speeds of 50, 75, 100 and 200 bauds, the WTK transmission system has a wide range of applications; it is suitable not only for transmitting telegraph messages and data over shortwave radio links, but can also be advantageously used over cables or radio relay links (without diversity operation), for example for linking to radio stations (Fig. 54). The principal application is the grouping of data channels for transmission over shortwave radio links. Another use is the simultaneous radio transmission on a single-sideband basis of several messages from stations at various locations in a radio network.

The operating principle of a WTK channel, consisting of modulator and demodulator and the necessary line matching module for interfacing to the transmission section is represented in Fig. 55.

5.2 Operating Principle of a WTK Channel

With the aid of a trigger stage the **modulator** converts the neutral or polar current signals offered to the input into steep-sided pulses. These drive an FM oscillator, which puts out the low and high shift frequencies for stop- and start-polarity, respectively. The subsequent filter clips the spectrum of the modulation products arising during keying, in order to minimize crosstalk to adjacent channels. The stopband attenuation of the filter for the adjacent shift frequency is 35 dB or better.

At the receive side, the aggregate signal passes through the matching transformer and an amplifier of the FLE to the **demodulator.** From the

84

incoming aggregate signal the receive filter picks out the frequencies allocated to its channel. The stopband attenuation of the receive filter is considerably higher than that of the transmission filter and is at least 60 dB. The purpose of the filters is to cut out crosstalk from adjacent channels, especially during severe selective fading. Via an amplifier/limiter, the signals are fed with constant amplitude to the discriminator. Level fluctuations due to fading of up to 40 dB are thereby compensated. The recovered d.c. signals are converted by a flip-flop into steep-sided pulses.

The **line-matching module** adapts the WTK channels to the input of the radio transmitter or output of the radio receiver, or to a long-distance line. It contains the line amplifiers and matching transformers necessary for d.c. decoupling of the transmitting and receive equipment.

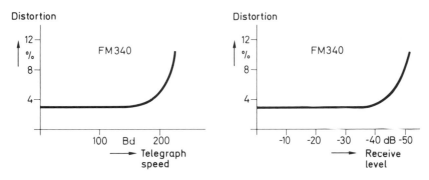

Fig. 56 Distortion in a WTK channel as a function of telegraph speed and receive level

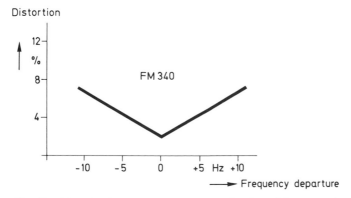

Fig. 57 Distortion in a WTK channel as a function of frequency departure

85

Characteristics of a WTK channel

The characteristics of a WTK channel as regards the quality of the transmitted information can be represented by the signal distortion curve. Figure 56 shows the distortion as a function of the telegraph speed and receive level. In Fig. 57 the distortion is plotted against the frequency deviation of the radio channel; over a WTK channel the distortion is less than 3%. The quality of a WTK transmission channel can also be assessed with reference to the bit error rate as a function of signal-to-noise ratio (see also Fig. 47).

5.3 Increasing the Transmission Reliability with WTK

The transmission reliability is increased by using the frequency diversity (Fig. 58) or space diversity method (Fig. 59). As will be shown later, the amplitude of two signals arriving over two different radio paths is evaluated, and the regenerated signals are passed on to the subscriber. A combined evaluation and summation circuit adds the d.c. signals of the two channels if the levels are equal or their difference is slight. If the levels differ by > 8 dB the channel with the weaker signal is cut off. Adding the two channels brings an increase in signal-to-noise ratio that is particularly important with low receive levels.

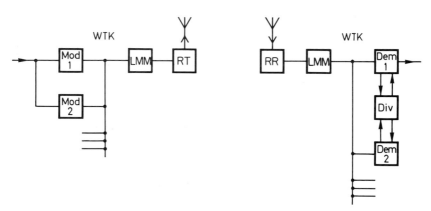

Fig. 58   WTK system with frequency diversity operation

Mod	1, 2 Modulator	Div	Diversity module
Dem	1, 2 Demodulator	RT	Radio transmitter
LMM	Line matching module	RR	Radio receiver

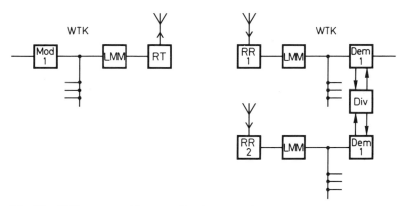

Fig. 59   WTK system with space diversity operation

On the other hand, if the level difference between the two channels is large, the signal with the low level and small signal-to-noise ratio will have a very adverse affect on the transmission quality. To avoid this, signals that are much weaker than the best-received signal are suppressed.

When a voice-grade channel is loaded with two frequency-diversity channels, the CCIR recommends [9] a channel frequency spacing of at least 400 Hz (400 Hz corresponds to a delay time difference of 2.5 ms).

5.4   Superimposed Telegraphy in Connection with WTK

Superimposed telegraphy permits speech and telegraphy to be transmitted simultaneously in the 300–3000 Hz voice band of a shortwave radio circuit. For this purpose the upper portion of the voice band is cut off by means of a frequency separating filter and made available for the transmission of telegraph messages. The speech signals are transmitted via the lowpass filter and the telegraph signals through the highpass network. The layout of a frequency separating filter for 4-wire operation is shown in Fig. 60.

This limitation of the speech band does not entail any appreciable degradation of logatom articulation and/or affect the timbre. With the speech cut off at 2.5 kHz, as is usually the case in practical systems, the logatom articulation still comes up to about 94%.

In 4-wire operation one independent channel is available for speech and for telegraphy in both directions of transmission. The two directions must be

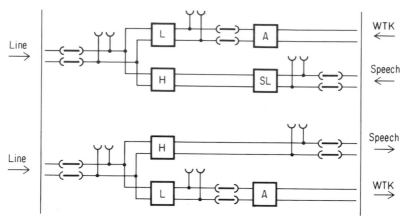

Fig. 60   Layout of a 4-wire frequency separating filter

L   Lowpass filter
H   Highpass filter
SL   Speech limiter
A   Attenuator

decoupled and the 4-wire separating filter assemblies therefore comprise one highpass and one lowpass filter in each direction of transmission. The lowpass filters intended for the send direction include speech limiters to avoid overloading of the transmission path. The attenuation characteristic of a 4-wire filter is shown in Fig. 61.

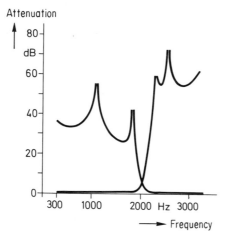

Fig. 61   Attenuation characteristic of a 4-wire frequency separating filter

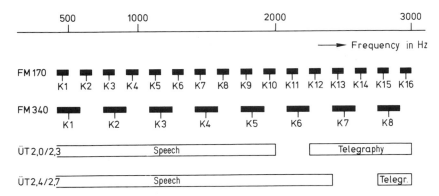

Fig. 62 Frequency allocation scheme for speech and telegraph transmission (K 1, K 2 ...
number of channels)

Only one frequency band is available for both directions of transmission in 2-wire operation. Different channel frequencies must therefore be used for transmitting the telegraph messages in the two directions.

The lower portion of the voice band may then be loaded with WTK channels of the FM 170 system to transmit one or two telegraph messages at a speed of 100 bauds.

The frequency separating filter UT 2.4/2.7 kHz (for limiting the speech band at 2.5 kHz) provides a telephone channel ranging up to 2.5 kHz. Telegraph messages may then be transmitted using channels 16 and 17 of the FM 170 system of the WTK 1000 terminal. The upper portion of the voice band may be used to accommodate from 1 to 4 WTK channels, depending on the type of frequency separating filter employed (Fig. 62).

6 Diversity Methods

The purpose of diversity methods is to reduce the disadvantages of shortwave radio transmission that result from different types of variation in the transmission characteristics of the radio link, especially fading. The fundamental principle of all diversity methods is that the message is recovered at the receiving end on a plural basis (multiple reception) and the quality of the transmission is upgraded by combination or selection of the signals. In this way, the mean signal-to-noise ratio is better and the error rate lower than with a single transmission path. This purpose can also be served by multiple transmission of the message on the send side. The waves arriving at the receiving station differ frequently in amplitude, phase and polarization. There are various means of multiple transmission; the methods discussed here are diversity procedures such as **space, frequency, antenna, time** and **polarization diversity.**

With single transmission of the message, space, antenna or polarization diversity can be employed. With the frequency and time diversity methods, the message is transmitted in duplicate.

The criteria for signal interpretation on the receive side are orientated according to the amplitude of the received signals. Automatic selection circuits always seek out the 'strongest' signal. The efficacy of the diversity methods can be further substantially increased by deriving evaluation criteria also from the coding equipment of the data protection terminals and by evaluating the distortion of the received signals. It is then also possible to diminish the influence of noise signals.

There should be a minimum of correlation between the various shortwave paths, i.e. the levels of the individual communications should fluctuate independently of each other, for evaluation according to intensity, if the diversity method is to be effectively applied. All shortwave links via the sky-wave are subject to selective fading on a very narrow band, which may lead to complete deletion of the signal (Fig. 67).

The ground wave, with its short range, is not affected by selective fading in the daytime. In the fringe area between sky- and ground waves, fading is also encountered, the effect of which can be reduced by diversity operation.

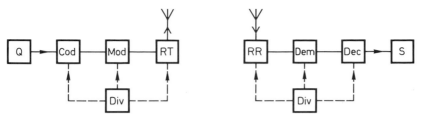

Fig. 63   Block diagram of a radio link with possible diversity equipment

Q	Source	Dec	Decoder	RT	Radio transmitter
S	Sink	Mod	Modulator	RR	Radio receiver
Cod	Coder	Dem	Demodulator	Div	Diversity equipment

In view of the variety of diversity procedures and equipment, Fig. 63 indicates the points in a radio link where diversity equipment may be applied. In the following, the diversity methods and evaluation procedure are described.

6.1 Frequency Diversity

With frequency diversity (Fig. 64) the same message is transmitted simultaneously by two radio transmitters tuned to different high frequencies (e.g. frequency spacing 3 MHz) and received by two separate receivers. Due to the great expense involved, this method is seldom used. It has its use, however, in cases where no-break communication is required during frequency changeover (transfer from day to night frequency).

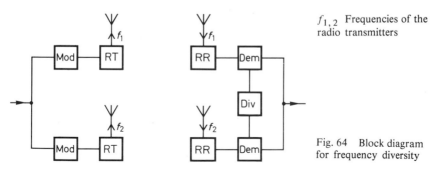

$f_{1,2}$ Frequencies of the radio transmitters

Fig. 64   Block diagram for frequency diversity

91

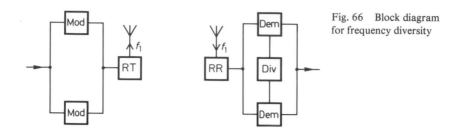

Fig. 66   Block diagram for frequency diversity

Demodulated signal, path 1

Demodulated signal, path 2

Mutilated signal

Resulting signal

Fig. 66   Graphic representation of two signals – receive paths 1 and 2 – and the resultant signal when using the diversity method

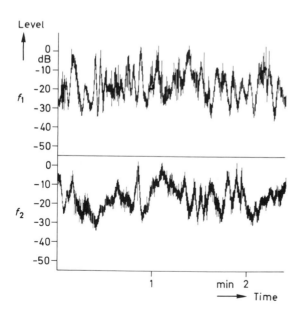

Fig. 67   Levels of two signal frequencies spaced by more than 400 Hz

92

When using voice-frequency telegraph systems for shortwave operation WTK (Chapter 5) two channels in the voice band of a single-sideband circuit carry the same message and are transmitted simultaneously by one radio transmitter (Fig. 65). To receive such transmissions, a single-sideband radio receiver is generally required. The most favourable WTK channel is selected in such a way that the signals of the two channels are added in order to improve the signal-to-noise ratio when the receive voltage difference is only slight. If the level difference is large (>8 dB) the channel with the weaker signal is cut off. The cost in terms of equipment, with this method, is low. The advantages of this method are that only one antenna and one radio receiver are required. This also means that frequency-diversity systems are very important for mobile applications, where space considerations generally preclude the setting up of more than one antenna. The spectrum requirements are doubled with frequency diversity since two channels are made available for one message.

Figures 66 and 67 represent the demodulated signals of two WTK channels. The levels of the two channels do not fluctuate synchronously. The channel frequencies are spaced by $\geqslant 400$ Hz. Figure 67 shows the level curve of the two frequencies over a period of several minutes. At no time are the signal amplitudes down simultaneously.

6.2 Space Diversity

The space diversity method − also known as the equipment diversity method − has acquired significance exclusively in non-mobile applications. This method uses **two** radio receivers with antennae of the same design, but placed some distance apart, to receive the same transmission (Fig. 68). In the two demodulators of the receiving system the signals are demodulated separately, then compared in an evaluating circuit and

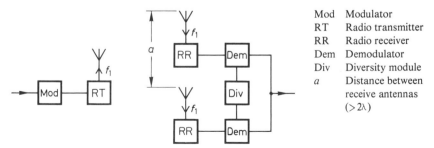

Mod	Modulator
RT	Radio transmitter
RR	Radio receiver
Dem	Demodulator
Div	Diversity module
a	Distance between receive antennas ($>2\lambda$)

Fig. 68   Block diagram for space diversity operation

93

Fig. 69   Block diagram for antenna diversity operation

combined, if both levels are good, to obtain a resultant signal. If the level
difference is greater than 8 dB, only the channel with the stronger signal is
connected through to the output of the demodulator. The efficacy of the
space diversity method depends on the distance between the antennae.
Depending on the wavelength of the desired transmitting station, a receiver
antenna spacing of about 2λ usually brings about a considerable
improvement.

6.3   Antenna Diversity

The antenna diversity method (Fig. 69) is a variant of the space diversity
method, whereby only **one** radio receiver is used, this being connected via
an antenna selection circuit usually to one of two antennae spaced some
distance apart. The selection circuit links the receiver to whichever antenna
is located in the area with the better field strength. The IF voltage is used as
the criterion for this. The point of switchover is determined by a pre-set
threshold value.

With the antenna diversity method less apparatus is required, but at least
two suitable antennae are required in order to effectively improve reception.

Antenna diversity is less efficient than frequency or space diversity because
a threshold must be manually set and use of the actually most favourable
antenna cannot be ensured.

6.4   Polarization Diversity

For polarization diversity (Fig. 70) two antennae are used, of which one is
e.g. polarized in the horizontal and one in the vertical plane. As with the

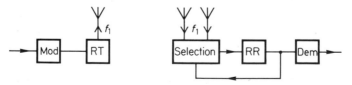

Fig. 70   Block diagram for polarization diversity operation

94

antenna diversity method, the antenna with a sufficiently strong receive signal is selected in each case. The antennae occupy little space and therefore lend themselves particularly well to mobile service.

6.5 Time Diversity

The time-diversity method (Fig. 71) involves transmission of the same signal twice with an interval of several 100 ms. Storage facilities are then required at the transmitting and the receiving end. Only one radio transmission channel is used. At the receive side, the signal transmitted first is compared with the delayed signal for information content. The intact signal is selected and fed into the terminal. Due to the predominantly burst-type noise on shortwave links, a considerable improvement in the quality of the transmission can be achieved by a suitable choice of the interval between the two signals. Since time diversity implies double transmission of the signals, the transmission capacity is halved. To have the same customer data rate on the receive side as on the send side, the signalling rate on the transmission path must be doubled.

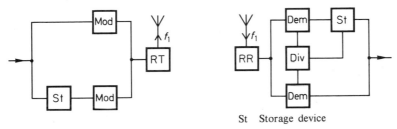

St   Storage device

Fig. 71   Block diagram for time diversity operation

6.6 Interpretation Procedures for Space and Frequency Diversity Operation

There now follows a short survey of the interpretation procedures, some of which have already been described in previous sections. The evaluation criterion is in most cases the reception level. By simple means, the demodulated signals can be added (**combination method**). It is also possible to evaluate the receive levels in the demodulators, and with the aid of this criterion to select the signal with the greater amplitude (**selection method**). The simultaneous use of both methods has become prevalent. Here, the signals are combined if the difference in their levels is small; this increases the signal-to-noise ratio. With level differences of >8 dB the receive channel with the weaker signal is cut off.

95

The functioning of the evaluation circuit can be disturbed by extraneous noise sources. Space diversity systems operating with one frequency are subject to equal interference on the different paths; frequency-diversity systems when evaluating the highest receive voltage may interpret a noise source as the 'best' signal. In order to avoid this, it is possible to use evaluation criteria derived from the amplitude and direct from the information content. The coded information from the data protection equipment can be used for this purpose (see Chapter 7).

6.6.1 Combination Method

With the combination method, the demodulated signal voltages put out by two receivers are added together. The aggregate voltage is fed to the output circuit (Fig. 72).

In FM transmission the amplitude of the signal voltage remains almost constant within the limitation range of the demodulator. However, there is a shift in the distribution of the useful and the noise components of the signal as the signal-to-noise ratio of the HF signal changes. From the point of view of circuitry, the combination method is easily implemented; namely, the signals are fed to a trigger circuit. If the signal-to-noise ratios of two demodulated signals do not differ by more than 8 dB, combination produces an increase in signal-to-noise ratio, if the noise power in the two channels is assumed to be equal and not correlated. If the two signals are equal in amplitude, the gain in signal-to-noise ratio is 3 dB.

This gain – which cannot be achieved with selection method – is decisive if the two signals each have such a small signal-to-noise ratio that undisturbed individual reception would no longer be possible.

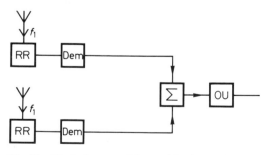

OU Output circuit
Σ Adder

Fig. 72   Block diagram of the combination method

Combination produces bad results if one of the channels has a substantially lower signal-to-noise ratio than the other, i.e. in the worst case supplies noise and interference only, for example. The assumption of equal noise power at the outputs of the two channels is then no longer justified, especially with frequency diversity. Nonetheless, diversity reception by the combination method brings a considerable advantage over single reception, since it is rare for one of the channels to be extremely bad.

6.6.2 Selection Method

With the selection method (Fig. 73) the information, after evaluation of the signals in both demodulators, is fed to the output circuit via the selection circuit. An electronic switch permits the commutation time to be kept short (e.g. 20 ms) in relation to telegraph pulse transmission.

In order to choose the demodulator supplying the 'best' signal, the demodulator signals are compared with each other for amplitude at a suitable point in the circuit (EV). The demodulator supplying the highest voltage is selected and connected to the output circuit.

The voltages compared constitute the sum of the signal and the noise voltage. The transfer characteristics of the demodulators must be equal before evaluation.

The selection method is effective when one of the two demodulators supplies a signal with a large signal-to-noise ratio. If, on the other hand, the signals of both demodulators have a small signal-to-noise ratio, both of the evaluated signals will also be bad. The selection circuit does not increase the signal-to-noise ratio, as is the case with signal combining.

EV Evaluation circuit
SE Selection circuit

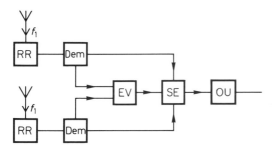

Fig. 73   Block diagram of the selection method

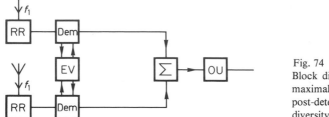

Fig. 74
Block diagram of the
maximal ratio
post-detection
diversity-combining method

6.6.3 Maximal Ratio Post-detection Diversity Combination Method

The addition of the demodulated signal voltages is also characteristic of this method, with the difference that the signals are evaluated before being combined (Fig. 74). Evaluation is performed, for example, by the demodulators. If both voltages are approximately equal in amplitude, both demodulated signals are used in their entirety for combining. If the compared voltages are dissimilar, on the other hand, the demodulator whose signal lies below the stronger received signal by more than 8 dB is cut off.

The demodulator whose signal level is less than 8 dB below the stronger received level contributes to the aggregate signal, with diminished output voltage.

The maximal ratio post-detection diversity combination method avoids the limitations involved in the combination and selection methods and is superior to either of these methods.

6.7 Efficacy of Diversity Methods

In practice, all the above-described methods are used. In commercial radio communications, the space- and frequency-diversity methods are chiefly used. Whereas the highly effective space-diversity method requires two receiving systems and large antenna installations and is therefore used in stationary applications, the frequency-diversity method must be given preference for mobile applications.

Diversity methods generally permit reduction of the error rate by a factor of 10 to 100. Figure 75 represents the error rate for single and diversity reception.

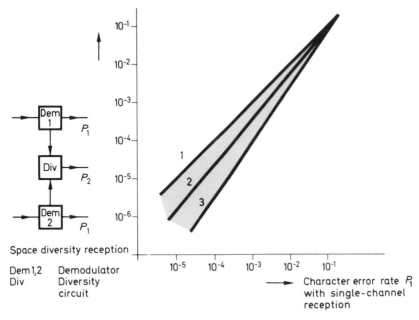

Fig. 75   Character error rate with two-antenna diversity reception and space diversity reception with two receivers compared with the error rate for single reception

Dem 1, 2	Demodulator	Curve 1	Single reception
Div	Diversity circuit	Curve 2	Two-antenna diversity
		Curve 3	Space diversity

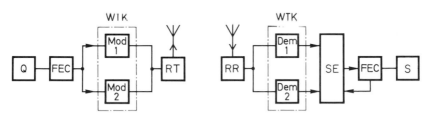

Fig. 76   Shortwave radio link for one-way traffic; the selection circuit for the diversity channels is controlled by the data protection system

Q	Source (e.g. teleprinter)	RT	Radio transmitter
S	Sink (e.g. teleprinter)	RR	Radio receiver
FEC	Data protection system	SE	Selection circuit
	(forward error correction)	Mod 1,2	Modulators
WTK	Voice-frequency telegraph system	Dem 1, 2	Demodulators

The evaluation criterion for the diversity methods considered so far is the level in the individual channels. The effectiveness of this procedure is badly impaired, for example, when extraneous signal sources interfere with one of the channels and make up a greater share of the aggregate level than does the useful signal, so that false evaluation is the result. By using data protection methods in connection with diversity channels the efficacy of the latter can be considerably increased, since distortion and code criteria assist the evaluation of the information. Interference with a channel by spurious signals can be effectively detected and combated by this method. Figure 76 represents the principle of a radio link with two channels operated in accordance with the frequency-diversity principle. The signals of the two channels are evaluated by a selection circuit which ensures an optimal decision in the selection of the channels by code and distortion evaluation.

7 Data Protection

In spite of modern modulation and demodulation methods and diversity reception in telegraph signal and data transmission over shortwave radio links the error rate on the transmission circuit may still be higher than expected. Recourse is then made to data protection systems. The transmission of encrypted messages over shortwave radio circuits requires a particularly high degree of reliability which can only be obtained with data protection systems. The latter use redundant codes to enable the receiving station to discover a falsification of the transmitted information. The protection methods differ as regards the code adopted and the procedure used for automatic error detection and correction. Operational requirements also have a bearing on the method to be employed.

7.1 Error Detecting and Error Correcting Codes

A distinction is made between error detecting and error correcting codes. Both are redundant codes.

The redundancy of a code is a measure of the extent to which it is capable of detecting or correcting errors. To achieve redundancy, the actual digital information is expanded by one or by several check bits (parity bits), which are formed in accordance with certain rules.

Redundant codes are, for instance, the constant-ratio code ITA No. 3 (3:4 ratio) as well as the parity check code (6 + 1) which lend themselves in particular to the detection of transmission errors on ARQ circuits. These two codes produce comparable results as regards error detection capability.

An error detecting code alone, however, does not permit error correction in practical operation unless the detection of an error is not at least indicated, e.g. by printout of an error symbol. To be really useful in practical applications, error detection must be accompanied by the 'automatic repeat request' feature which causes the errored part of the message to be repeated until it is found to have been correctly transmitted.

Protection methods employing error detecting codes, automatic repeat request and repetition in the event of a detected error are referred to as repeat-on-request methods (ARQ methods).

Error correcting codes are not restricted to mere detection of errors. They are rather supposed to permit the correction of detected errors at the receiving end. In view of the enhanced requirements, the error correcting code will of necessity include more redundancy than an error detecting code.

In the case of error correcting codes the system operator depends on their error detecting and correcting power and under extremely adverse transmission conditions the incoming message must be accepted as it is when a repeat transmission cannot be initiated.

Protection methods based on the use of error correcting codes are classified as forward error correcting codes (FEC principle). FEC systems using convolutional codes with bit-by-bit coding have performed well in shortwave radio circuits.

Data protection systems may be divided into error detecting and error correcting systems. Error detecting systems failed to gain any significance worth speaking of. By contrast, error correcting systems have given very satisfactory service. These may be classed into systems based on repetition request in connection with an error detecting code and systems employing an error correcting code and the forward error correcting principle. A comparison between error detecting and error correcting systems is shown in Fig. 77.

7.2 Redundancy

The term redundancy implies 'superfluity' and, in the communication engineering field, 'extra information'. Redundancy may be achieved, e.g. by transmitting extra bits in addition to the information bits, i.e. bits not actually required for information interchange, in an effort to protect the information from transmission errors. In view of the fixed relation with the actual information, this redundancy permits a check to be made at any time as to whether the information has been correctly transmitted.

The redundancy is generated in the coder of the transmitting station prior to transmission. It is derived from the message proper and transmitted together with the message. It enables the receiving station to verify correct reception. In the event of an errored transmission, a correction of the error can be initiated. An error is detected in the receiving station by new code words being generated which fail to agree with the prearranged coding.

This advantage of verifying correct reception is traded off against an

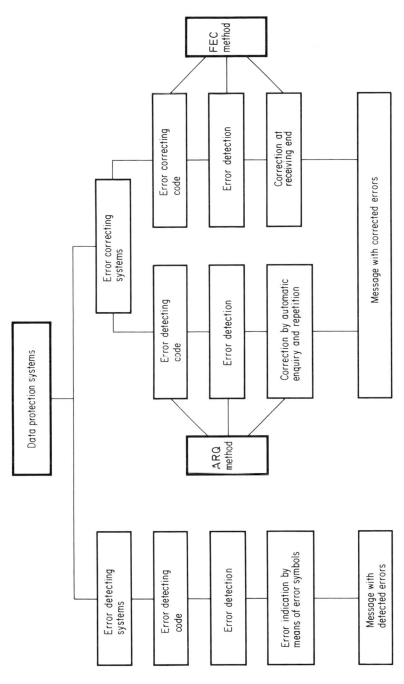

Fig. 77 Comparison between error detecting and error correcting systems for protected data transmission

103

increase in the volume of information to be transmitted. In other words, either the bandwidth of the transmission channel must be increased or a longer transmission time must be accepted.

In spite of redundancy, some errors may go undetected during message transmission. Where a particularly high degree of security is required, the redundancy must be expanded accordingly. Error control is, however, restricted by practicability considerations. A residual error rate must be tolerated for every transmission. To determine the error control method most suitable for a given application, it is indispensable to establish first of all the error profile of the transmission path to permit optimal adaptation of the redundancy.

7.3 Hamming Distance

In connection with data protection, the Hamming distance provides a measure of the error detecting capability of a code. This measure indicates the number of digit positions that must be different for any two characters belonging to the same code. This is equivalent to the minimum number of bits that must be falsified to produce a valid, although undetected incorrect character.

With a Hamming distance of $d = 1$, every mutilated character is changed into an adjacent character, belonging to the same code. In this case the code is free from redundancy but does not permit error detection or error correction. A generally known example is the 5-unit code (ITA No. 2) where the falsification of only one bit will produce another valid character so that the error cannot be detected.

If the Hamming distance is $d = 2$, all those mutilations can be discovered which affect an odd number of bits of the mutilated character. Automatic repetition request may here be adopted. However, correction without repetition of the mutilated character is not possible. A well-known example is the constant-ratio code ITA No. 3, a 7-unit code featuring a fixed ratio between stop polarity and start polarity bits. It permits also even-numbered bit falsifications to be detected in those cases where no transpositions have occurred.

Correction by reconstruction of the received signal without the aid of a backward channel for repetition request is possible only with a Hamming distance of $d \geqq 3$. The range that can be economically used extends from $d = 3$ to $d = 5$. As can be seen, the Hamming distance is a decisive factor for the error detection capability of a code.

7.4 Protection Methods

Dependent on the type of shortwave link a distinction is made between the following protection methods:

▶ Protection methods for duplex connections, employing an error-detecting code, automatic repetition request and signal repetition over a continuously available backward channel (ARQ system). Messages are exchanged simultaneously in both directions on a point-to-point transmission basis.

▶ Protection methods for simplex connections, again with the aid of an error-detecting code. Every time a message has been transmitted, the connection is interrupted and an automatic turn-around effected so that correct reception can be acknowledged. Whenever signals are mutilated as a result of interference, the distant station is requested to repeat the message. Messages can be sent in only one direction at a time and the circuit must be 'turned around' for transmission in the opposite direction.

▶ Protection methods for one-way traffic with no backward channel being required for error correction. An error-correcting code is used in this case which permits transmission errors to be continuously corrected immediately at the receiving station. This method is referred to as Forward Error Correction (FEC). Messages can be transmitted in one direction only to any number of recipients. Therefore, these protection methods lend themselves particularly well to broadcast traffic, e.g. for maritime and meteorological services as well as for press, police and embassy networks.

On duplex circuits data are protected by means of ARQ systems which permit simultaneous traffic flow in both directions and repeat falsified portions of the message over the backward channel in response to an automatic repetition request. On board a ship or in the case of mobile radio stations, i.e. in all those cases where duplex traffic can hardly be realized because radio transmitter and radio receiver, together with their antennae, are located close together, the message being transmitted would impair reception of the message sent from the distant station. A simplex ARQ system is used here to protect the data on the transmission path and suitable radio transceivers for rapid changeover between transmission and reception are required which are controlled by the simplex ARQ system. In this way up and down working is possible, using only one radio frequency and one antenna.

The first two of the previously mentioned protection methods are based on the repetition request principle. They are intended for point-to-point traffic with an acknowledgement being sent from the data sink to the data source. A big advantage of an ARQ system is the fact that the transmitting end is always aware of the extent to which reliable message transmission to the distant station can be achieved.

This ARQ method cannot be used on radio circuits which do not provide a backward channel. Forward error correction must here be adopted for error control.

Protection systems using forward error correction for one-way traffic provide no indication at the transmitting end whether or not the message has been successfully transferred to the receiving end. These systems must therefore be adapted to the error mode of the shortwave link so that most of the transmission errors, caused by fading or other transient disturbances, can be corrected.

Whereas the start–stop principle is employed for transmitting characters over unprotected shortwave radio links, as on physical telegraph circuits, data protection systems used on radio links depend on the synchronous transmission principle [16]. This reduces their proneness to interference.

In the case of start–stop working with the 5-unit code a start element is placed at the beginning of each character which causes the teleprinter at the receiving end to start up. The five information-bearing elements of the character are followed by a stop element, mostly 150% of the unit pulse length, which causes the machine to stop. Assuming a nearly equal cycle speed of the transmitting and the receiving machine, the required synchronism is ensured for the duration of one character. If a start element is mutilated on the transmission path, the receiving machine will fail to detect it and misinterpret any code element, having start pulse polarity, as the start element so that synchronism is lost. A mutilated start element will therefore frequently produce a series of falsified characters. In a synchronous system, on the other hand, only one character will be affected by a falsified signal element.

7.5 ARQ Method

ARQ systems permit the protected transmission of enciphered or plain-text telegraph messages and data over shortwave radio links. The principle of automatic error correction by repeating signals that had been falsified (ARQ method according to van Duuren) constitutes the most efficient method of correcting detected transmission errors. It ensures a degree of reliability even on noise-corrupted links which may be compared to that obtained, for instance, on cable circuits or radio relay connections.

The ARQ principle includes

▶ detection of errored characters

▶ request for and execution of a repeat transmission of the characters found to be errored (automatic repetition request and repetition), and

▶ calldown of characters from the subscriber station only during non-cycling periods and provisions to prevent the output of signals as long as the circuit is disturbed.

For the purpose of error detection, **redundancy** is incorporated in the 5-unit code character and in our case it is expanded by two elements so that a 7-unit code signal results. This code enables the ARQ receiver to detect signals falsified *en route* and to initiate repeat transmission of the signals last transmitted. In contrast to the subscriber side, where the characters are called down individually in start—stop mode, the 7-unit code signals are transmitted in the synchronous mode on the radio path. This necessitates synchronism between the transmitting and receiving facilities of the inter-operating stations. Bit and character synchronism are automatically established and maintained by crystal oscillators and synchronizing circuits.

The fundamentals of automatic error correction by signal repetition are shown in Fig. 78. The receivers in both stations A and B check all incoming signals for the correct polarity ratio established by the 7-unit code. If a falsified character has been detected, in station B for instance, during transmission from A to B, the first action taken is the blocking of message output to the subscriber. Transmission in the opposite direction from B to A is also suspended. Instead, a repetition request in the form of the RQ signal is sent to station A. Reception of this signal in station A causes the current transmission to be interrupted and the repetition process to be tripped. Station A transmits first the RQ signal, this time to be interpreted as a confirmation signal, to station B. This is followed by the repeat transmission of the three signals last transmitted to station B (four-character repetition cycle). These three signals are kept in storage in anticipation of a possible request for retransmission. The last three signals that were sent in the opposite direction from B to A are also repeated, no matter whether one of these signals was falsified or not. The repetition cycle

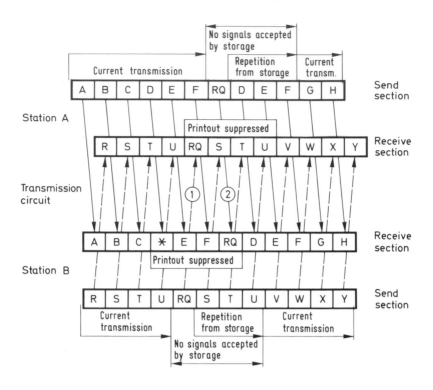

Fig. 78   Repeat transmission with four-character repetition cycle

A to Y	Transmitted and received characters	RQ	RQ signal
*	Character found to be errored in the receive branch of the ARQ equipment	①	Repetition request signal
		②	Confirmation signal

can be extended to eight distributor cycles (8-character repetition cycle), including one RQ-signal and seven stored characters, to account for the delay on very long transmission circuits.

Signal repetition in both directions continues until the confirmation RQ has been correctly received in station B 'gated RQ'. To enhance transmission reliability further, the 'tested RQ' mode of operation may be employed. In this case all characters included in a repetition cycle are checked and a character found to be wrong will initiate a further repetition cycle at the end of a current cycle.

If the retransmitted characters have been correctly received in station B, the repetition cycle is terminated. These characters are then passed on to the receiving device which records them in proper sequence and immediately after the previously received message text.

Full-duplex transmission circuits are the precondition for automatic repetition request and error correction. Figure 79 shows the layout of a transmission circuit equipped with ARQ systems. Each ARQ system contains the circuit units for the transmitting and the receiving legs. Of the two inter-operating protection systems one is conditioned to work as the 'master' station, station A for instance, and the other as the 'slave' station.

The transmitting branch of the ARQ system calls the message down from the data terminal equipment on a character-by-character basis, converts the code used by these terminals, e.g. the 5-unit code, into an error-detecting code, e.g. the 7-unit code and stores the signals last transmitted in anticipation of a repetition request.

The receive branch examines the signals arriving via the radio receiver for transmission errors and initiates a repetition cycle upon the detection of an

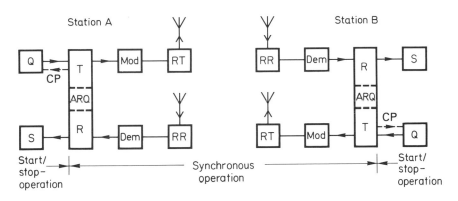

Fig. 79 Shortwave radio link employing the ARQ data protection system

Q	Source	Mod	Modulator
S	Sink	Dem	Demodulator
ARQ	Data protection system	RT	Radio transmitter
T	Transmit branch	RR	Radio receiver
R	Receive branch	CP	Call-down pulse

Fig. 80 Teleprinter 1000 for call-down operation in connection with ARQ terminals

errored signal. It reconverts the error-detecting code into a code acceptable to the data terminal equipment, adds the start and stop elements, and feeds the correctly received signals to the data sink.

The ARQ clock ensures synchronous signal transmission and supplies all the timing pulses required for controlling all functions of the ARQ system.

The duration of the repetition cycle depends on the signal delay on the transmission path. If the four-character repetition cycle (one RQ signal and three repeated characters) proves to be too short on a circuit with an extremely long signal delay (e.g. on cascaded radio and cable circuits, see Section 10.3, p. 182), an eight-character repetition cycle may be employed which comprises one RQ signal and seven repeated characters.

In the subscriber stations message flow is suspended for the duration of the repetition cycle. No signals are picked up from the data source and no signals are transferred to the data sink. Figure 80 shows a teleprinter 1000 which permits output of characters by means of call-down pulses. Teleprinters of this type are used in connection with ARQ terminals.

In some cases ARQ systems are used for one-way transmission. The repetition feature has to be deactivated for these applications. Any character, e.g. the Space combination, may be chosen as an error sign to mark falsified characters.

For data transmission with automatic error correction single-channel terminals (Section 7.5.1 and 7.8) and multichannel terminals (Section 7.5.2) are available.

110

7.5.1 Single-channel ARQ System

For shortwave radio links carrying a low traffic volume a single-channel connection will be adequate. Data transmission in this case is protected by single-channel ARQ terminals.

Code employed for error detection

In telegraph traffic the Alphabet ITA No. 2 is used, each character of this 5-unit code being represented by a combination of five information bearing elements. A total of $2^5 = 32$ code combinations can be formed, which are all assigned to the characters of the telegraph alphabet. In addition, the steady-start criteria occurring during the transmission intervals, i.e. steady-start polarity and steady-stop polarity, must be transmitted. The ARQ system encodes these criteria and transmits them as special idle signals. Steady-start polarity and steady-stop polarity are transmitted in the form of idle signals alpha and beta. Apart from these two signals, automatic error correction depends on a further code combination, the RQ signal. To be able to accommodate these additional code combinations, the 5-unit code must be expanded by an additional signal element, the identification bit. This bit precedes the five-code bits. As a result, the character set has been extended by another 32 code combinations of which, however, only three are actually used. To permit a distinction between the 32 characters of the telegraph alphabet and the three additional 'function' signals, the identification bits of the latter are assigned a different polarity.

The 5-unit code signal, already supplemented by the identification bit, is expanded once more, this time by a parity bit which is added as the last bit to the code combination. The 7-unit code thereby obtained makes it possible to detect signals falsified on the transmission path. This 7-unit code affords $2^7 = 128$ possible combinations, of which 35 are used. The structure of the 7-unit code is shown in Fig. 81.

1st signal element (identification bit)

This is a start-polarity element for the 32 characters of the 5-unit code and a stop-polarity element for idle signals α and β and for the RQ signal.

2nd to 6th signal element (5-unit code groups)

These signal elements are those of the 5-unit code characters to be transmitted (ITA No. 2).

	A	B	C	D	E	F	G	H	I	J	K	L	M	N	O	P	Q	R	S	T	U	V	W	X	Y	Z	<	≡	A…	1…	IS	RQ	α	β
LTRS	–	?	:	✛	3	∎	∎	∎	8	♫	(	)	.	,	9	0	1	4	'	5	7	=	2	/	6	+	<	≡	A…	1…	IS	RQ	α	β
FIGS	A	B	C	D	E	F	G	H	I	J	K	L	M	N	O	P	Q	R	S	T	U	V	W	X	Y	Z	<	≡	A…	1…	IS	RQ	α	β
5-unit code group (ITA No.2) — 1	●	●		●	●	●				●	●						●		●		●		●	●	●	●			●	●				
2	●		●				●		●	●	●	●				●	●	●			●	●	●					●	●	●				
3			●			●		●	●		●		●	●		●	●		●		●	●		●	●				●		●			
4		●	●	●		●	●			●	●		●	●	●			●				●		●			●		●	●				
5		●					●	●				●	●		●	●	●			●		●	●	●	●	●			●	●				
Identification bit — 1																																●	●	●
5-unit code group — 2	●	●		●	●	●				●	●						●		●		●		●	●	●	●			●	●				
3	●		●				●		●	●	●	●				●	●	●			●	●	●					●	●	●				
4			●			●		●	●		●		●	●		●	●		●		●	●		●	●				●		●			
5		●	●	●		●	●			●	●		●	●	●			●				●		●			●		●	●				
6		●					●	●				●	●		●	●	●			●		●	●	●	●	●			●	●				
Parity bit — 7	●			●				●	●		●	●		●	●		●	●	●			●		●		●				●				

Fig. 81   5-unit code according to the Alphabet ITA No. 2 and 7-unit code employed by the ARQ single-channel terminal

□   Start polarity   ∎   Free for domestic use but not permitted for international traffic
◉   Stop polarity
<   Carriage return   ✛   Who are you?
≡   Line feed   ♫   Bell

7th signal element (parity bit)

The polarity of the parity bit is chosen so that the sum of the stop-polarity elements will always be an odd number. If an even number of stop-polarity elements results for the first six signal elements, a stop-polarity element will be added as the seventh element. On the other hand, if the number of the stop-polarity elements is an odd number, a start-polarity element will be added.

Parity, code, and element tests are performed by the single-channel equipment for each individual character to detect any errors and to request signal repetition. With the aid of the parity bit each character is checked at the receiving end as to whether the stop-polarity elements within a character result in an odd number. If interference causes several signal elements to be falsified in such a way that the parity requirement is satisfied (transposition errors), the error detection function is taken over by the code tester and the element tester.

Expansion of the 5-unit code to six signal elements results in $2^6 = 64$ code combinations of which 32, as previously mentioned, are used for communication purposes while three are used as special function signals.

112

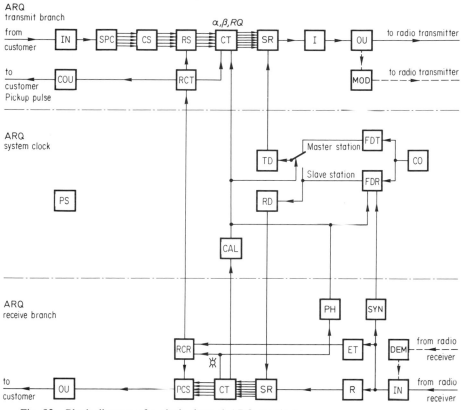

Fig. 82　Block diagram of a single-channel ARQ terminal

ARQ	Transmit Branch	CT	Code translator
IN	Input circuit	RCR	Repetition cycle timer, receive-side
SPC	Series–parallel converter	PCS	Parallel–series converter
CS	Character storage	OU	Output circuit
RS	Repetition storage	COU	Calldown output circuit
OU	Output circuit		
RCT	Repetition cycle timer, transmit-side	ARQ	Clocking System
CT	Code translator	CO	Crystal oscillator
SR	Shift register	FDT	Frequency divider, transmit-side
I	Polarity inverter	FDR	Frequency divider, receive-side
MOD	Modulator	TD	Transmitting distributor
		RD	Receiving distributor
ARQ	Receive Branch	CAL	Calling facility
DEM	Demodulator	PS	Power supply
IN	Input circuit		
SYN	Synchronizing circuit	RQ	RQ signal
R	Polarity restorer	⌭	inadmissible code combination (errored character)
ET	Element tester		
PH	Phasing circuit	α	Idle signal for continuous start polarity
SR	Shift register	β	Idle signal for continuous stop polarity

113

Since the remaining 29 code combinations are not required for message transmission and therefore are not sent out by the transmitting branch, it is possible for the receiving branch to identify these code combinations as being inadmissible.

In addition to the parity and code tests an element test is provided to determine the degree of distortion of the individual elements of a character. Interference along the transmission path, as is known, causes variations in the duration of the signal elements or, in other words, the signals are distorted. If the distortion within a character exceeds the $\pm 25\%$ mark, the element tester will reject this character.

Automatic repetition request will be initiated by the single-channel ARQ terminal whenever a code combination other than one of the 35 admissible code combinations is received (code or parity error) or the distortion limit is exceeded.

Operating principle of a single-channel ARQ terminal

The function of a single-channel ARQ terminal will be explained with the aid of block diagram Fig. 82. Each terminal station contains the circuit units for the transmit and receive branches as well as the common clock. As previously mentioned, two ARQ terminals are interconnected so that one works as the master station and the other as the slave station. A single-channel ARQ terminal is shown in Fig. 83.

Fig. 83   Single-channel ARQ terminal ARQ 1000 duplex in carrying case

ARQ transmit branch

The message signals to be transmitted are clocked out of the data terminal in series configuration and applied to the transmit branch by way of input circuit IN. The series–parallel converter SPC, which is next in the circuit, converts the serial characters into parallel characters, and these are transferred to the 1000-character storage CS. The storage always acts as a buffer storage when the message source is not designed for controlled signal pickup. When the storage is filled to 95% of its capacity, a lamp warns against further data input in order to prevent overflow of the storage.

From the character storage CS the message is accepted into the following repetition storage RS, provided that the two interconnected stations are in phase and no repetition cycle is in progress. The repetition storage stores three or seven characters as required for the selected repetition cycle. The subsequent code translator CT checks the polarity of the first six signal elements and selects the seventh signal element (parity bit) so that there is an odd number of stop polarity pulses. Continuous start or stop polarity conditions at the input are translated by the CT into the idle signal α or β, while a repetition request signal RQ is placed ahead of a repetition cycle. The send side repetition cycle timer RCT is designed to block the emission of pickup pulses while a repetition cycle is in progress to suppress the transfer of characters from the buffer storage to the repetition storage, and to transmit the three or seven characters stored in the repetition storage. Shift register SR converts the parallel signals into serial signals for transmission over the radio link.

To enhance the reliability of automatic phasing at the receiving end, the polarity of every first 7-unit code character is reversed in polarity inverter I (marked cycle). The message is then applied to the output circuit OU, where it is converted into neutral-current or polar-current dc signals and passed on to the radio transmitter. With single-sideband transmission, the modulator MOD of the WTK 1000 system is used to generate the frequency-modulated signals.

ARQ receive branch

The 7-unit code signals from the radio receiver are fed to input circuit IN of the ARQ receive branch. The frequency-modulated signals of a WTK single-sideband transmission are demodulated by the WTK demodulator DEM and then applied to input circuit IN. The phase relationship of the signal elements and signals is checked in synchronizing circuit SYN and phasing circuit PH and corrected, if necessary. At the same time element tester ET checks whether the signal elements are distorted and, if the value of $\pm 25\%$ is exceeded, initiates a repetition cycle. Polarity restorer R cancels the cyclical polarity inversion on the send side. After the serial signals have been converted into parallel signals in shift register SR, a code and parity check is performed in code translator CT. If an errored signal ☖ – number of signal elements with stop polarity is even – or the RQ signal is detected, the receive-side repetition cycle timer RCR delivers a criterion to the send-side repetition cycle timer RCT, tripping a repetition process. If idle signal α or β is received, the parallel–series converter PCS receives control signals which cause the transmission of steady start or steady stop polarity to the customer station. The parallel–series converter changes the parallel signals back into serial signals and, with the start and stop elements added, transmits them to the customer station via output circuit OU in the form of neutral-current or polar-current signals.

ARQ system clock

The system clock supplies the pulses required to control the ARQ system. A frequency of 3686.4 kHz with a maximum drift of $\pm 1 \times 10^{-6}$ is generated in a crystal oscillator CO. For the desired transmission rates, adjustable in steps between 50 and 2400 bauds, this frequency is scaled down by frequency dividers FDT and FDR for the ARQ transmit and the receive branch, respectively. From this frequency the distributors TP and RD derive the clock pulses for the transmit and receive legs which control and sequence the functions in the single-channel ARQ terminal. In the slave station the relevant clock pulses for the transmitting distributor TD and the receiving distributor RD are derived jointly from the frequency divider FDR. The power supply PS provides the supply voltages for the whole ARQ terminal.

Calling facility

A calling facility CAL permits either station to activate or deactivate a point-to-point circuit which is in the standby condition. The call procedure is initiated by depressing a button on the ARQ terminal. The radio transmitter associated with the calling station then switches on automatically. The radio transmitter in the called station also comes on when the address has been received and an acknowledgement signal is subsequently returned. After the calling station has received the acknowledgement signal it sends β idle signals or the address. These signals are used to synchronize the called station. After synchronization, transmission can begin. If no connection is set up despite there being a call, the calling station will automatically return to the standby (ready-to-receive) condition.

A connection can be cleared down by either station by transmitting an end-of-call signal.

Enhanced-protection circuit for crypto operation

Additional protective measures have been taken to reduce the occurrence of transposition errors during the changeover from idle signals to text and vice versa. Moreover, the RQ signal is sent twice. The enhanced-protection circuit is of particular importance for crypto operation since these transpositions or the loss of an RQ signal may result in the loss of crypto synchronism. When this feature is used, the 4-character repetition cycle is replaced by a 5-character one.

7.5.2 ARQ Multichannel System

Time-division multiplexing in conjunction with automatic repetition request is employed to permit a number of independent and protected messages to be transmitted simultaneously over a shortwave radio link. Two or four channels are here combined to achieve economical multiple utilization of the transmission path. The messages of the individual channels are inter-leaved. A two-channel system comprising channels A and B is referred to as **diplex.** Figure 84 shows a diplex connection for two channels in each direction of transmission. Two diplexes of this type can be combined so as to obtain a four-channel time-division multiplex system, the two additional channels being designated as channels C and D. Time-division ARQ terminals are capable of interoperating on an international basis, the code employed for error detection and the error correcting procedure being specified by the applicable CCIR Recommendations [4, 17] for universal application. To permit their connection to telex networks, these ARQ TDM systems may be equipped with facilities intended to control switching functions [14].

The channels afforded by an ARQ multichannel system may be subdivided

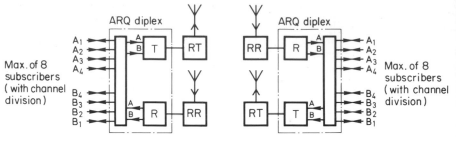

Fig. 84   Radio circuit employing two-channel ARQ data-protection terminal and main channel division

RT	Radio transmitter	A1–A4 ⎫	Subscriber inputs and outputs
RR	Radio receiver	B1–B4 ⎭	with channel division
A	A-channel	T	Transmit branch
B	B-channel	R	Receive branch

in a simple manner for the purpose of **sub-channel operation,** i.e. a main channel is split into subchannels operating at a reduced transmission rate. This allows subscribers with a low traffic volume to share the advantages of protected transmission at fees lower than those of a main channel. Postal administrations, for instance, use sub-channels to handle their telegram traffic or lease them for point-to-point communication.

Apart from international telex traffic, two-channel data protection terminals are now increasingly used in small and medium-size private communication networks.

Transmission of enciphered messages over shortwave radio links makes the use of data protection systems mandatory. Otherwise, i.e. without the use of data protection equipment, interference along the transmission path may cause a loss of the cipher phase (Section 10.2, p. 178) resulting in the loss of the entire message. Multichannel ARQ systems are therefore provided with special facilities which additionally protect the transitions from idle signal to message signal and vice versa as well as the repetition request signal RQ because a simulated transition or a falsified RQ signal will also result in the loss of the cipher phase. To this end a repetition is initiated with each transition between text signals and idle signals to determine whether this was an actual transition or only a signal mutilation. Moreover, double transmission of the RQ signal ensures that signal repetition proceeds reliably and correctly.

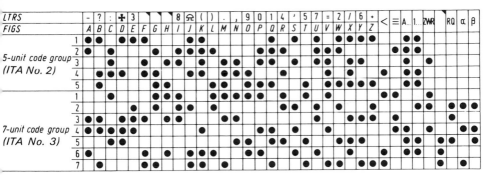

Fig. 85   Code table of the 5-unit code ITA No. 2 and of the 7-unit code employed by the ARQ system

□	Start polarity (−)	◪	Free for domestic use but not permitted for international traffic
◉	Stop polarity (+)		
<	Carriage return	✚	Who are you?
≡	Line feed	♫	Bell

Code employed for error detection

A 7-unit binary code is again used for error detection, which affords a total of $2^7 = 128$ code combinations. Out of this character set 35 code combinations are used for a 3:4 constant-ratio code, each code combination consisting of three stop-polarity elements and four start-polarity elements. Thirty-two code combinations are assigned to the characters of the telegraph alphabet ITA No. 2 (Fig. 85). The remaining three combinations are required to represent the idle signals α and β as well as the RQ signal. If a signal with a polarity ratio other than 3:4 is received, the ARQ system will interpret this as a falsified signal and repetition of this signal will be triggered automatically. Figure 86 shows the repetition cycle in a two-channel time-division multiplex terminal employing a four-character repetition cycle.

If an equal number of start-polarity and stop-polarity elements have been falsified within a character due to signal **transposition** *en route,* the 3:4 polarity ratio is retained and a wrong character will be printed. The probability of transposition errors is very low, however, and as an additional safeguard the individual elements of each signal received are checked for their distortion (element test). A distortion exceeding a preset value of e.g. ±25% will also cause signal repetition.

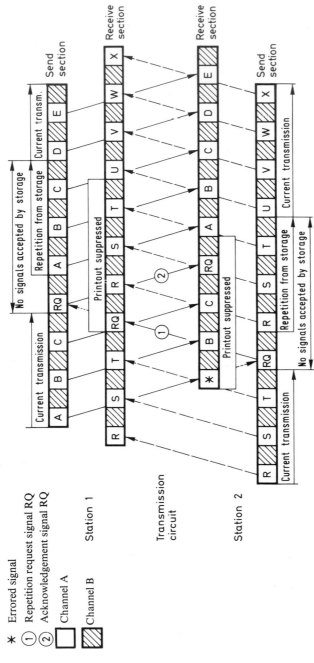

Fig. 86   Repetition process of a two-channel time-division multiplex terminal employing the four-character repetition cycle

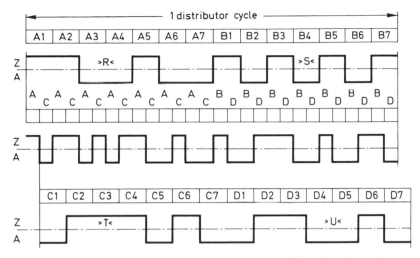

Fig. 87   Interleaving of signal elements with two-channel and four-channel operation

Channel A: 'R'   Channel C: 'T' (inverted)
Channel B: 'S' (inverted)   Channel D: 'U'

Time-division multiplex principle

ARQ data protection systems may be designed as multichannel systems for two or four channels, dependent on the transmission capacity to be provided by the radio link. In the case of four-channel operation the messages of every two channels A and B as well as C and D are element-interleaved. Figure 87 shows signal element interleaving with two-channel and four-channel operation. To permit a clear distinction between the individual channels A and B as well as C and D in four-channel operation, channels B and C are transmitted with inverted polarity in accordance with a CCIR Recommendation [4].

Channel division

Telegraph and data channels operated over radio links with automatic error correction in continental and intercontinental traffic are not only used for telex service by the postal administrations and private carriers but are frequently leased for point-to-point connections which are available at any time without the need for dialling the distant station. Many subscribers insist on the possibility of transmitting messages immediately, i.e. without

121

the delay involved by connection set-up. On the other hand, their message volume is such that paying the rental fees of a full channel could not be justified. For economical reasons it will therefore be advisable to subdivide the full channel into time slots. A **channel divider** is therefore provided which permits each channel to be split up into two, three or four subchannels. The transmission capacity of such a subchannel amounts to 1/4, 1/2 or 3/4 of the full channel capacity. The message transmission time increases by a factor of two, three or four but the costs will proportionately decrease, compared with those of the full channel. The message signals are picked up from the subscriber terminals in rotation by the channel divider. With a main channel split up into two subchannels, for instance, the characters of the two channels are transmitted in alternate succession. At the receiving end they are distributed to the two subscribers again in alternate succession. During the period one subscriber receives a signal, steady-stop polarity is sent to the other subscriber. The baud rate as such, 50 bauds for instance, remains unchanged but the transmission capacity actually available to the subscriber is halved.

To permit automatic phasing at the receiving end also in the case of subchannel operation, every fourth character of a channel is marked by

Table 11   Main channel dividing options of an ARQ multichannel terminal

Division ratio	Number of subscribers	Share	Subchannel sequencing			
1/1 (one main channel)	1	1/1				
2 × 1/2 (two subchannels)	1	1/2	1		3	
	2	1/2		2		4
1/4 + 3/4 (two subchannels)	1	1/4	1			
	2	3/4		2	3	4
2 × 1/4 + 1/2 (three subchannels)	1	1/4	1			
	2	1/4			3	
	3	1/2		2		4
4 × 1/4 (four subchannels)	1	1/4	1			
	2	1/4		2		
	3	1/4			3	
	4	1/4				4

inverting its polarity. The dividing options for a main channel are shown in Table 11 together with the interleaving of the characters and their distribution to the subchannels.

Telex service over shortwave radio links

ARQ systems with automatic error correction by signal repetition on the transmission circuits paved the way for the present worldwide telex network operating on shortwave links. The first intercontinental telex circuits had been established in 1950 between Europe and the USA with the aid of electromechanical ARQ terminals.

Today more than one million subscribers are connected to the international telex network and most of them are provided with the facilities for dialling themselves the desired connection over international telegraph circuits. This applies in particular to those cases where connections are not or no longer established over shortwave links. A great many international telex links are implemented by means of cables (submarine cables) or, during the last few years and to an ever growing extent, by means of satellite links. These links are not beset by the difficulties encountered in shortwave operation, such as fading, so that special measures need not be taken to improve transmission reliability.

However, where international telex connections have to be operated over shortwave radio links, a special switchboard must be interposed. Although ARQ terminals are capable of ensuring an adequate freedom from errors in the received messages, message flow is interrupted during interference periods, i.e. during signal repetition. In the case of self-dialling service this would leave the customer in doubt, in particular in the event of protracted interference, as to which parts of his message have already been sent out. He might even be led to believe that the connection no longer exists. On the other hand, if a connection is set up by a switchboard operator in the outgoing country, the connection can be monitored and the subscriber may be informed of its status, if necessary. If certain conditions can be met, outgoing international traffic may be handled on an automatic basis.

Incoming international traffic is in most cases automatically routed to the domestic network. The selection information for setting up a telex connection over ARQ-protected shortwave links must be offered to the ARQ terminal in the 5-unit code. In national telex networks employing number-plate dialling the selection information must therefore be converted into keyboard selection signals. In the outgoing direction the selection information is generated manually by the switchboard operator on his teleprinter keyboard. The incoming keyboarded selection signals are converted into dial signals in the destination country by means of selection signal translators.

For transmitting messages in the telex network over ARQ protected connections, each message must be buffered before it is transmitted. The interim storage may be assigned to the switching facilities or to the ARQ terminal. The ARQ terminal calls the message in from the storage character-by-character by means of tripping pulses.

For the described telex connection both the go channel and the return channel of an ARQ circuit are required for setting up the connection. A telex connection established in this way will therefore not permit simultaneous transmission in both directions. Even though it functions as a duplex connection on the radio path, it can only be used for conversational traffic because the teleprinter of the subscriber is only designed for half-duplex traffic, i.e. for one-way-at-a-time transmission.

The ARQ multichannel terminal, used on all international telex connections established over shortwave links, supplies to the telex centre all the signalling criteria required for establishing and clearing down a connection.

Counters report the transmission or reception of a predetermined number of idle signals α or β which are used as circuit seizure and release signals. The charges per call for the aforementioned international connections are in many cases recorded by call time meters actuated by the operator of the international switchboard. Timing pulses sent by the ARQ terminal at intervals of 6 or 10 seconds increment the call time meter during interference-free periods. Signal repetition periods are, of course, not registered. In large telegraph offices equipped with extensive international switchboards, call data recording for international calls may be accomplished automatically by centralized metering facilities [15]. The chargeable time may also be made a function of a certain degree of circuit efficiency (e.g. greater than 80%). In ARQ terminals the efficiency is ascertained by registering the repetition cycles or counting the number of the signals correctly transmitted (Section 7.5.4). A low efficiency, i.e. reduction of the message flow due to interference on the transmission path, causes the connection to be taken down. The subscriber can be informed accordingly by the exchange. The call must then be booked once more by the subscriber at a later time and set up again by the operator of the international telex switchboard.

As a rule, the outgoing country must adapt its signalling system to that of the destination country. A restriction must be made, however, for telex traffic over ARQ links. Special facilities are provided here to adapt the connection set-up procedure to the signalling system employed for ARQ links [14].

Conversion of the signalling criteria must be accomplished in the respective country both for the outgoing and the incoming traffic. If the outgoing country employs signalling system B (CCITT U1), for instance, it must take care of the conversion from CCITT U1 into U20. If the incoming country employs type-A signalling, the CCITT U20 signalling criteria are converted into U1, type-B criteria.

An example of an international telex connection of a country is given in Fig. 88.

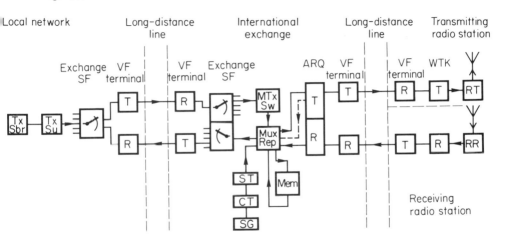

Fig. 88 Layout of an international telex connection established over an ARQ system (semi-automatic traffic, i.e. manual handling of outgoing calls and incoming automatic dialling)

TxSbr	Telex subscriber	Mem	Memory
TxSU	Telex signalling unit	ST	Selection signal translator
SF	Switching facility	CT	Code transmitter
VF	Voice-frequency telegraph transmission terminal	SG	Signal generator
		ARQ	ARQ terminal
T	Transmit branch	WTK	Voice-frequency telegraph transmission terminal for operation over shortwave links
R	Receive branch		
MuxRep	Mux relay repeater		
MTxSw	Mux-telex international switchboard	RT	Radio transmitter
		RR	Radio receiver

125

Operating principle of an ARQ multichannel terminal

The function of an ARQ multichannel terminal will now be explained with the aid of the block diagram in Fig. 89. Figure 90 shows the front view of such a terminal. This diagram shows a two-channel terminal (diplex) with the associated subchannel facilities. The channels are given the designations A and B. In the case of four-channel operation a second diplex is added, the additional channels being the channels C and D. Each diplex contains the circuitry for the transmitting and receive legs as well as the common clock. Two terminals inter-operate over the shortwave radio link in the synchronous mode, one being conditioned as master station and the other as slave station.

ARQ transmitting branch

The serial message signals arrive via input circuit IN at the series–parallel converter SPC which converts them into parallel signals and applies them to the delay line DL. The individual characters are clocked out of the data terminal equipment by the tripping pulse circuit TPC. The messages may also be transferred in parallel configuration to the ARQ terminal. Delay line DL together with the series–parallel converter SPC permits the messages to be buffered for a delay time between data terminal equipment and the transmission terminal amounting to the length of two characters. The signals are then passed on to the repetition storage RS where they are kept in anticipation of a request for repetition. Following this the signals of channels A and B are, in the case of two-channel operation, switched alternately to code translator CT which translates the five code elements into 7-unit code signals consisting of three stop-polarity elements and four start-polarity elements. The output shift register OSR reconverts the parallel signals into serial signals.

With four-channel operation the messages of channels A and B are element-interleaved with those of channels C and D. To mark the channels, the polarity of the signals is periodically inverted (polarity inverter I).

At the output the transmitting branch supplies, via output circuit OU and tone modulator TM, d.c. or VF signals by way of the conventional telegraph or data interfaces for the connection of transmission facilities or for the direct connection to a radio transmitter.

126

ARQ receiving branch

Arriving from the radio receiver, the 7-unit code signals reach the receive branch by way of input circuit IN or the tone demodulator TDM. Polarity inverter I restores the signals previously inverted by the transmit branch of the distant terminal.

The input shift register checks the received signals for the 3 : 4 polarity ratio. A signal found to be correct is changed back to the 5-unit code by code translator CT and fed on to the parallel–series converter PSC of the associated channel. Having been supplemented with the start and stop elements, the signal is passed on to the subscriber via output circuit OU.

An element tester ET checks whether or not the signal elements exceed the admissible degree of distortion. If a signal is found to be disturbed, a repetition cycle is initiated.

No signals are put out to the subscriber line during cycling periods. Instead, steady-stop polarity current is sent to the subscriber. The repetition cycle is controlled by the receiving end and transmitting end repetition cycle timers RCR and RCT, respectively.

In the case of channel subdivision each subscriber is assigned input and output circuits and a series–parallel converter at the send side. The channel dividers CD call the signals down periodically from the subscriber terminals and feed them on to the delay line DL.

Clocking system for the ARQ terminal

The frequency clock FC generates a frequency of high stability from which the required timing and sequencing pulses are derived by frequency division. One frequency divider FDT and FDR, respectively, is provided for the transmitting branch and the receiving branch so that the receive branch pulses can be displaced in the master station with relation to the transmit branch pulses depending on the signal delay on the transmission path. In the slave station the frequency dividers of the transmit branch and of the receive branch are interlocked. The element synchronizer SYN ensures that the elements of the individual signals are always sampled in their centre. The character synchronizer PH permits automatic phasing of the ARQ terminal to achieve the correct character phase.

7.5.3 Efficacy of ARQ systems

Experience has shown that the error rate may be expected to decrease by a factor of from 100 to 1000 on the different types of radio links under

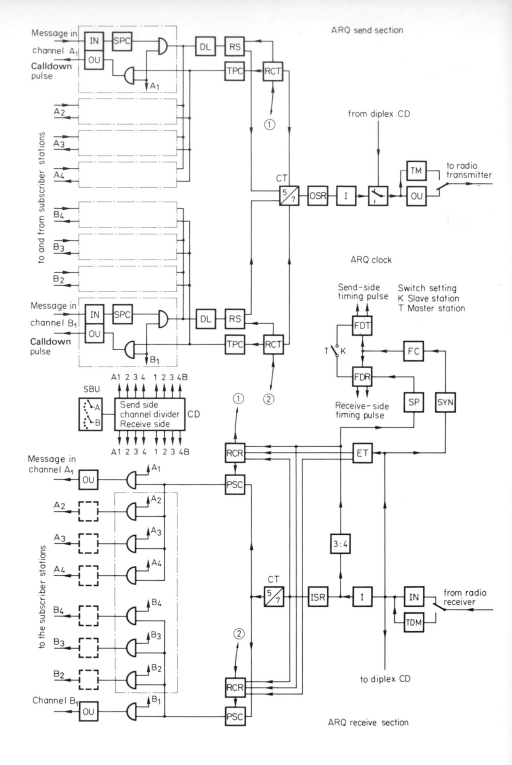

Fig. 90   Two-channel ARQ
multiplex terminal ELMUX 1000
in carrying case

◄   Fig. 89   Block diagram of an ARQ multichannel terminal

TPC	Calldown pulse circuit	SPC	Series–parallel converter
SP	Automatic phasing circuit	SYN	Element synchronizer
OU	Output circuit	TDM	Tone demodulator
OSR	Output shift register	FC	Frequency clock
CT	Code translator	TM	Tone modulator
IN	Input circuit	I	Polarity inverter
ISR	Input shift register	RS	Repetition storage
FDR	Frequency divider, receive side	RCR	Repetition cycle timer,
FDT	Frequency divider, transmit side		receive side
CD	Channel divider	RCT	Repetition cycle timer,
PSC	Parallel–series converter		transmit side
SBU	Switch module for	3 : 4	3 : 4 polarity check
	subchannel operation	A1, B1	Main channels
DL	Delay line	A2–A4	Subchannels
ET	Element tester	B2–B4	Subchannels

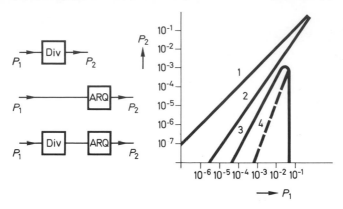

Fig. 91   Residual error rate if data protection and diversity reception are employed

P_1 Error rate on transmission path
P_2 Error rate at the output of the transmission terminals (residual error rate)
Curve 1   Transmission without data protection and without diversity reception
Curve 2   Transmission with diversity reception
Curve 3   Transmission with ARQ data transmission system
Curve 4   Transmission with ARQ data transmission system and with diversity reception

normal operating conditions when error-correcting systems depending on signal repetition are used, compared with unprotected radio links (see Table 17). The remaining errors are the result of transpositions, i.e. of disturbances causing an equal number of start and stop polarity elements to be falsified. The efficacy of ARQ terminals is shown in Fig. 91. It is possible to determine from this figure the residual error rate of a shortwave radio link. The curves represent the mean values. The improvement achieved with diversity reception is indicated to permit a comparison. Depending on the configuration of the diversity facilities, the error rate can be reduced by a factor of from 10 to 100. The broken line shows the residual error rate to be expected with diversity reception and data protection.

Signal repetition caused by a transmission error causes the message flow to be interrupted. The efficiency of a transmission terminal decreases with the number of repetition cycles. If the error rate on the transmission path reaches a very high figure P_1 equalling about 3×10^{-2}, the message flow is interrupted by the continual signal repetitions.

The efficacy of the 7-unit code employed in ARQ terminals compared to the 5-unit code used in start-stop equipment can be determined arithmetically

130

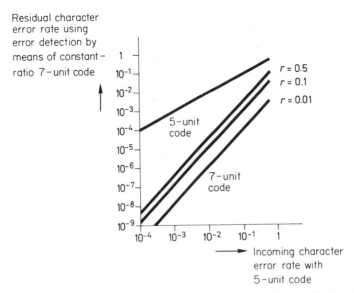

Fig. 92 Residual error rate with a random distribution of the mutilations of the 7-unit code signals protected by error detection compared with 5-unit code signals without error detection (r = bias factor)

and is shown in Fig. 92 for disturbances having a random distribution. In digital signal transmission the individual bits of a binary signal may be differently affected so that start-polarity bits, for instance, are affected more than stop polarity bits. The reason for this is the fact that selective fading and selective interferers may impair one of the two shift frequencies in the case of frequency shift keying more than the other frequency for the duration of several bits or characters. This leads to bias mutilation within the characters and bias factor r has been introduced to define this condition. It states the number of mutilations of one bit polarity in relation to the total number of disturbances. Disturbances which happen to transform a start-polarity bit into a stop-polarity bit, as previously mentioned, and at the same time a stop-polarity bit into a start-polarity bit are called transpositions. They occur very infrequently. Assuming a random distribution of interference, the probability of a falsification of an equal amount of start- and stop-polarity bits is greatest, the bias factor r being 0.5. Where a bias condition prevails (factor r smaller than 0.5 but greater than zero), the number of these errors decreases considerably. From Fig. 92 it may be seen that under the most adverse conditions, symmetrical disturbances with bias factor 0.5, an error rate as low as 2.5×10^{-5} results for the 7-unit code while the error rate for the 5-unit code would be 1×10^{-2} under the same transmission conditions. It follows that the 7-unit code achieves a decrease in the error rate by a factor of 2.5×10^{-3}.

7.5.4 Degree of efficiency of an ARQ terminal

As is known, the message flow is temporarily interrupted in ARQ systems to repeat signals in the event of detected errors. The message throughput is thereby reduced. The degree of efficiency varies between zero (continuous repetitions) and 100%, depending on the error rate on the transmission path.

On a transmission circuit protected by ARQ systems the efficiency and, consequently, the quality of the connection can be continuously monitored with the aid of an efficiency meter. Figure 93 shows the efficiency as a function of the frequency of detected errors, assuming a uniform distribution of disturbances (the most unfavourable case so far as the degree of efficiency is concerned).

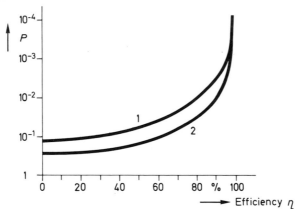

Fig. 93   Degree of efficiency of ARQ systems as a function of the error rate with four-character and eight-character repetition cycles

Curve 1   8-character repetition cycle
Curve 2   4-character repetition cycle

132

If the number of repetitions rises within a certain period, the number of the correctly transmitted signals decreases. If N is taken to be the number of characters that can be transmitted within a certain time and R the number of the repetitions that became necessary during the same period, the degree of efficiency η in per cents, assuming a four-character repetition cycle, will be

$$\eta = \frac{N - 4R}{N} \times 100\%$$

With two characters falsified within a message 100 characters long the degree of efficiency is

$$\eta = \frac{100 - (4 \times 2)}{100} \times 100\% = 92\%$$

In the case of ARQ terminals employing the eight-character repetition cycle the degree of efficiency will decrease correspondingly:

$$\eta = \frac{100 - (8 \times 2)}{100} \times 100\% = 84\%$$

Automatic billing in telex traffic over radio links is based on transmission time measurements. The result is indicative of the degree of efficiency of the transmission path. In the event of a disturbed radio link, continuous determination of the degree of efficiency is to prevent any substantial discrepancy between the charged transmission period and the time actually available for transmission. The degree of efficiency is ascertained by counting the number of the repetition cycles within predetermined periods. If this value drops below 80%, control signals may be sent to the connected switching facilities or to the subscriber, and signals derived for checking the transmission path and for taking the existing connection down. The connection will be re-established only when the degree of efficiency rises again above 80%. The required data for an efficiency meter are contained in a CCITT Recommendation [15]. Today ARQ terminals are equipped with these efficiency meters.

7.6 FEC System for One-way Radio Links

Systems employing an error-correcting code, permitting continuous error correction at the receiving end, protect telegraph messages on one-way shortwave radio links and in broadcast systems. They are referred to as

Fig. 94 Shortwave radio link for one-way traffic employing forward error correction (FEC)

FEC	Forward error correcting system	RR	Radio receiver
Q	Source	S	Sink
RT	Radio transmitter	Cod	Coder
		Dec	Decoder

Forward Error Correcting systems, abbreviated to FEC systems.* Errored signals can be corrected without a backward channel being available if codes having a greater redundancy are used (Fig. 94). If the use of forward error-correcting systems is planned on shortwave radio links, the error structure regarding frequency and temporal distribution of transmission errors along the route must be known before a decision can be reached on the required degree of redundancy and the type of code to be employed.

Correction with the aid of an error-correcting code

The forward error-correcting method here described employs a **convolutional** code which is largely adapted to the type of error structure found on the shortwave radio link, such as burst errors or random errors. A **diffused convolutional** code with an information bit/test bit ratio of 1 : 1 proved to be particularly suitable. A test bit is interposed between every two information bits I so that a parity bit or check bit CI can be derived with the aid of modulo-two-adders from a certain number of preceding and succeeding information bits. Since a test bit is transmitted with each information bit, the speed on the transmission path is twice the input speed. As far as the source is concerned, any input code may be used, as this is a **code-transparent** system.

At a telegraph speed of 50 bauds, the system is capable of completely correcting error bursts having a density up to 100% (i.e. all bits within the error burst are falsified) and a duration of up to 7 s if error-free bit sequences are received immediately ahead of and behind the error bursts in guard spaces amounting to three times the maximum length of the error bursts (21 s). In practice most error bursts result in a density of not more

---

* The designation FEC is occasionally also used for error detection with the aid of a time-diversity method.

134

than about 50% so that the full correction capability of the code is not exploited. In these cases the code also permits the correction of errors occurring during the intervals before and behind the error bursts. The lower the error burst density, the more errors can be corrected in the guard spaces.

Function of a coder/decoder

The error-correcting code used in the FEC system deals with the message on a bit-by-bit basis and is therefore directly suited to the coding of isochronous bit sequences, irrespective of the code employed. The characters are transmitted on a synchronous basis between coder and decoder. To permit operation with terminal equipment working on a start-stop basis and using the telegraph code ITA No. 2, asynchronous/ synchronous converters are required at the transmitting side and synchronous/asynchronous converters at the receiving end. Messages may be put in on the keyboard of a teleprinter or picked up character by character from a tape transmitter. In the serial characters furnished by the tape transmitter the stop element may be 50%-prolonged. In the case of a keyboarded transmission the stop element may have any length. Since both the coder and the decoder operate with character sequences of uniform bit duration, the telegraph characters as well as the steady start- and steady stop-polarity conditions are coded into an isochronous bit sequence. The five information-bearing elements of the telegraph code ITA No. 2 are translated into a 7-bit code with parity bit. The steady start- and stop-polarity conditions are also translated into 7-unit code combinations. The combinations used for this purpose are not assigned to telegraph characters and differ from the latter by the polarity of the first signal element.

Fig. 95   Forward Error Correcting System FEC 100 A containing both the coder and decoder

At the receiving end the parity bit of the 7-unit code permits the detection of uncorrected characters whenever the capability of the error-correcting system has been overtaxed as a result of abnormal interference along the transmission path.

In this case any character, 'Space' for instance, may be sent to the terminal equipment so that the number of the transmitted characters is retained. This is of importance in encrypted message transmission to maintain the cipher phase. Figure 95 shows a Forward Error Correcting System FEC 100A containing both the coding and the decoding circuitry.

Coder

The coder is shown in Fig. 96. The telegraph characters arriving from a terminal equipment via input circuit IN are fed to the asynchronous/synchronous converter (start-stop converter) SSC in the case of start-stop operation. The telegraph characters can be applied to the coder either in a direct way or by means of call-down pulses. The five information bearing elements are converted into parity-protected 7-unit code signals.

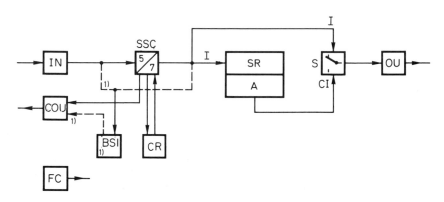

Fig. 96   Coder

FC	Frequency clock	CR	Call repeater
COU	Calldown pulse output circuit	SR	Shift register
OU	Output circuit	BS1	Bit synchronizer
IN	Input circuit	S	Switch
A	Adder	SSC	Start-stop converter
I	Information bits	(1)	for synchronous operation
CI	Checking information		

136

This asynchronous/synchronous converter is by-passed for coding messages from data sources supplying isochronous bit sequences. With synchronous operation, the coder will pick up the characters from the data source on a bit-by-bit basis. Where this is not possible, the internal timing can be adapted to the phase position of the incoming signals by means of a bit synchronizer BS 1.

Within the coder the incoming information bits I are fed through a multistage shift register SR. Information bit I, fed into the shift register at a certain time, is simultaneously passed on to the transmission circuit via switch S. At the same time a check bit CI is formed by certain, selected positions of the shift register through modulo-2 adder A, whose polarity depends on the checked storage cells. Switch S transfers and the check bit CI is sent to line via output circuit OU. With the next step of the shift register the next code bit with its associated check bit is fed out. To maintain the transmission speed for the subscribers, while the number of the check bits is the same as that of the code bits, the transmission rate must be doubled on the line. This is accomplished by switch S whose dwell time equals half the width of the code and check bits. The result is an isochronous signal at the coder output, alternate bits containing useful information and check information.

Decoder

The transmitted isochronous message signals arrive in the decoding circuit (Fig. 97) by way of input circuit IN. A bit synchronizer BS2 adapts the incoming message to the decoder timing. The individual signals of the bit stream are split up in code synchronizer CSY into code bits and check bits. Bit synchronizer and code synchronizer maintain the code phase also in the presence of disturbances on the transmission path lasting up to thirty minutes. As on the send side, the information bits I are passed through a shift register SR which has the same structure as the shift register in the coder. Employing the same rules, adder A forms the check information CI'. The transmitted check information and the check information formed in the decoder are compared bit by bit in comparator H. If the two check bits agree, the information has been transmitted without error. If the transmitted check information has been falsified *en route* or if the locally generated check information CI' does not agree with the transmitted check information CI, a corresponding error criterion is stored in syndrome register SR.

137

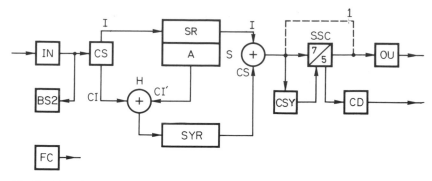

Fig. 97   Decoder

FC	Frequency clock	S	Selector switch
OU	Output circuit	CI, CI′	Checking information
CS	Code synchronizer	CD	Call decoder
IN	Input circuit	SR	Shift register
SYR	Syndrome register	BS2	Bit synchronizer
H	Half adder	SSC	Start-stop converter
I	Information bits	CSY	Character synchronizer
CS	Correction signal	(1)	for synchronous operation
A	Adder		

Depending on the distribution in time of errored information and check bits, an 'error pattern' results in the syndrome register. A decision logic determines from this error pattern whether the bit present at the output is falsified or correct. If a falsified information bit has been received, a correction signal from the syndrome register causes the polarity of this bit to be inverted by switch S, i.e. the bit is now corrected. The character synchronizer CSY establishes the character phase upon switching on the equipments, determining which of the seven bits of a current bit stream belong to a character. The isochronous 7-bit characters are recoded into the 5-unit configuration by the synchronous/start-stop converter SSC and, after adding the start and stop elements, passed on in bit-serial form to the terminal equipment via output circuit OU. Converter SSC also contains the logic for detecting a continuous stop polarity condition.

Timing

The clock pulses required for the coders and decoders are derived from a crystal oscillator FC by frequency division. The phase position of these dividers can be varied by synchronizers BS1 and BS2.

138

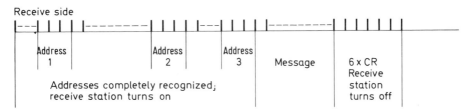

Fig. 98 Transmitting a call with the aid of the automatic call repeater

CR   Carriage return signal
LTRS   LTRS signal

Digital calling system

In a shortwave radio network any subscriber may call the desired distant station by transmitting a digital call. When a call has been detected, the respective terminal equipment turns on automatically. The calling procedure permits connections to be set up between subscribers on an individual call, group call, or broadcast basis. Using the proper address, individual subscribers can be called selectively or a group of subscribers can be called simultaneously.

The address is placed ahead of the message to be transmitted. As in the message proper, the individual signals of the address are telegraph characters. The call address consists of three characters out of the 26 possible combinations of the telegraph alphabet and an address-framing character (Fig. 98). To protect the address against transmission errors, the selection information is transmitted several times in succession. A call is interpreted as correctly received when at least three addresses have been correctly received within a certain period of time. The call address may be transmitted either from the keyboard of the terminal equipment or from a previously prepared perforated tape.

A call repeater CR (Fig. 96) may be used for a repeated automatic transmission of the address. The data source must then transmit the address ahead of the message. No message signals are accepted by the coder as long as calling signals are being transmitted. The call-down pulses are inhibited during this period by the call repeater.

At the receiving end the call decoder interprets the address and, after recognizing the associated address signals, completes a path to the terminal equipment. It disconnects the terminal equipment upon reception of the end-of-call signal. The beginning-of-call signals preceding the message are also used as end-of-call signals. The terminal equipment is cut off when several consecutive 'CR signals' have been received by the call decoder.

The calling procedure employing the automatic call repeater is shown in Fig. 98.

7.6.1 Efficacy of an FEC system

A knowledge of the possibility of reducing the error rate is of great importance for the use of data protection equipment on shortwave radio links. The improvement in message transmission by a protection system depends essentially on the correction capability of the system and on the frequency and distribution in time of the transmission errors. The bit error probability on the transmission path and the distribution in time of the bit errors, i.e. a knowledge of the error structure, are of particular importance in forward error correcting systems. As is generally known the efficiency is a measure of the quality of a connection in ARQ protection systems (see Section 7.5.4). To achieve an average efficiency of more than 80% and a character error probability of about 1×10^{-5} in the case of protection systems using automatic repetition request and signal repetition, the mean error probability of a transmission channel must not exceed a certain value. Similarly, forward error-correcting systems also require a certain minimum quality of the transmission channel if a satisfactory efficiency is to be ensured.

As mentioned in Chapter 1.9.3, shortwave radio links are affected chiefly by error bursts, in addition to random errors. Error-correcting codes employed with a forward error-correcting system should therefore be designed with emphasis being placed on error bursts.

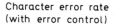

Character error rate
(with error control)

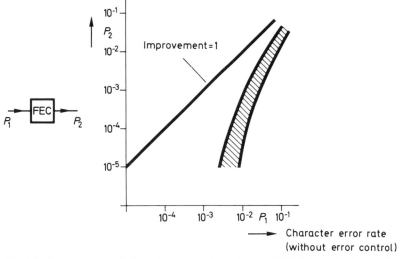

Fig. 99   Improvement of character error rate, using an FEC data protection equipment without backward channel

Intervals in %

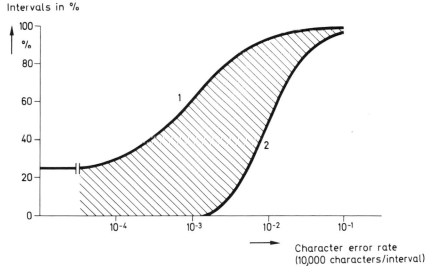

Fig. 100   Character error rate of a communication circuit with and without FEC data protection

Curve 1   Protected
Curve 2   Unprotected

On shortwave radio links with different distances and frequencies many measurements have been carried out with FEC-protected and unprotected transmission channels to determine the decrease in the character error rate achieved with FEC methods. As can be seen from the characteristics in Fig. 99, the character error rate dropped from 1×10^{-2} on the unprotected shortwave link to 4×10^{-5} after introduction of the FEC system. The improvement factor in this case amounted to 250. To provide further information on the statistical behaviour of practical transmission circuits, the character error rate for unprotected and protected transmission circuits is shown in Fig. 100. These curves reveal that the character error rate amounts to about 10^{-2} for 50% of the measuring intervals on unprotected circuits. When FEC systems are used, the character error rate dropped to about 6×10^{-4} for 50% of the measuring intervals. For 25% of the measuring intervals a character error rate of about 5×10^{-3} was obtained on unprotected circuits while errors had been completely corrected with the aid of an FEC system.

7.7 FEC System for One-way Radio Links with Channel Selection

The forward error correction system with channel selection (Fig. 101) is intended for use on excessively interference-prone shortwave radio links. It contains the same coding and decoding circuitry for error correction as the system described in Chapter 7.6. A selection process based on majority voting and quality evaluation (AMB and AQB, Fig. 104) offers the additional facility of selecting out of three channels, carrying the same message, the least disturbed channel. The correction capability of such a system is thereby appreciably improved. Figure 102 shows the front view of such a terminal.

Fig. 101   Shortwave radio link for one-way traffic with forward error correction (FEC) based on channel selection after quality detection

Q	Source	RR	Radio receiver
FEC	Forward error correcting system	Dem	Demodulator
Cod	Coder	CH	Channel selector
Mod	Modulator	Dec	Decoder
RT	Radio transmitter	S	Sink

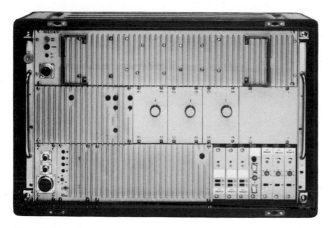

Fig. 102   Forward error correcting system with channel selection FEC 101 (in carrying case)

The system described employs an error correcting code which permits incoming errored messages to be continuously corrected at the receiving end. In the transmitting station the coder derives a check information from the message. Useful information and check information are transmitted in alternate succession by way of three WTK channels, each channel being delayed by 4 bits with respect to the neighbouring channel. The same message is therefore available three times at the receiving station. Since selective fading is known to affect only parts of the voice band at a given time, a channel selection circuit, which is based on majority voting and supported by quality evaluation, causes only the least disturbed message to be switched through to the decoder. The decoder detects falsified characters with the aid of the check information and initiates correction of the corrupted bits.

Coder

The coder shown in Fig. 103 converts the telegraph signals, arriving in the form of ITA No. 2 code signals, in its start-stop converter SSC into 7-bit code signals which are subsequently applied in the form of an isochronous character sequence I to the shift register. Messages may be transmitted in the start–stop mode with or without individual signal calldown. In the former case, there is no need for adjusting the calldown pulses COU so as to adapt them to the delay time of the subscriber station loop.

143

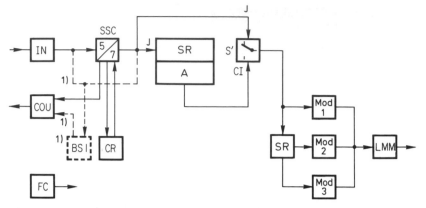

Fig. 103   Coder with three send channels

IN	Input circuit	SR	Shift register
COU	Calldown pulse output circuit	A	Adder
FC	Frequency clock	S	Switch
SSC	Start–stop converter	CI	Checking information
BS1	Bit synchronizer	Mod 1, 2, 3	Modulator
CR	Call repeater	LMM	Line matching module
I	Information bits	(1)	For synchronous operation

The information is fed through shift register SR. At certain points of this shift register adder A computes the check information CI. Switch S passes the actual message and the check information on to modulator Mod 1 and, delayed by 4 or 8 bits in shift register SR, to modulators Mod 2 and Mod 3. The information contained in the three channels is combined by line matching module LMM and forwarded to the radio transmitter. Convolutional coding results in an increase in the telegraph speed on the radio circuit.

If the source supplies isochronous character sequences, the latter are fed by input circuit IN immediately to the input of shift register SR, by-passing converter SSC. The bit synchronizer BS1 ensures correct timing of the sampling instants and, if the terminal equipment is clocked by the FEC, correctly timed emission of the calldown pulse to the source.

Decoder with channel selection

The information arriving over the radio path at the line matching module LMM is passed on to the three demodulators Dem 1, 2, 3 of the channel

144

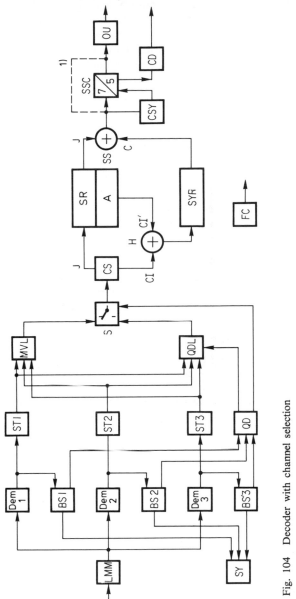

Fig. 104 Decoder with channel selection

LMM	Line matching module	I	Information bits	SR	Shift register
Dem 1, 2, 3	Demodulator	CI, CI′	Checking information	A	Adder
BS 1, 2, 3	Bit synchronizer	C	Correction signal	SS	Selector switch
SY	Sum synchronizer	SSC	Start-stop converter	SYR	Syndrome register
ST 1, 2, 3	Storage	CD	Call decoder	FC	Frequency clock
QD	Quality detection	S	Switch	CSY	Character synchronizer
MVL	Majority voting logic	CS	Code synchronizer	OU	Output circuit
QDL	Quality voting logic	H	Half adder	(1)	For synchronous operation

145

selection circuit (Fig. 104). The bit synchronizers BS 1, 2, 3 and the sum synchronizer SY are provided to establish synchronism between coder and decoder and to time sampling of the information. The latter passes storages ST1, ST2 and ST3 which have the additional function of cancelling the time displacement by 4 and 8 bits between the adjacent channels. The decision logic QD evaluates each channel for a period corresponding to the capacity of storage ST3 to determine whether the signals are to be selected on the basis of majority voting (MVL) or channel quality detection (QDL). In the majority voting logic MVL the signals are summed and subsequently applied to the decoder via switch S. At the same time, the decision logic QD checks each channel on the basis of various channel quality criteria. If the decision logic QD detects degradation of the signals in two of the three channels, the signal derived from the majority voting logic MVL is disconnected and the signal supplied by the channel quality detection logic QDL is connected to the decoder via switch S. This signal corresponds to the least disturbed channel at the time.

Message text I and check information CI are separated in the code synchronizer CS. As in the coder, message text I is fed through shift register SR to generate check information CI' in adder A. This check information is compared with the transmitted check information in half adder H. In the event of a discrepancy between CI and CI', an error indicator is stored in syndrome register SYR for the later correction of errored bits.

The contents of this register are evaluated and a correction signal C is computed. Falsified bits are corrected in selector switch S'S'.

Converter SSC recovers the original start–stop signals from the incoming isochronous 7-unit code bit stream, with character synchronizer CSY ensuring proper framing.

Digital calling unit

The digital calling unit permits the transmitting station to invite the receiving stations to receive on an individual call, group call, or broadcast basis. The source transmits the desired address to the coder where call repeater CR automatically repeats the address (Fig. 103). At the receiving end, the call decoder CD recognizes the associated character sequences (address) and puts the message through to the terminal equipment, turning the latter on at the same time. The address combinations can be set by means of switches provided on the calling unit.

7.7.1 Effectiveness of an FEC System with Channel Selection

In order to determine the effectiveness of an error corrector employing channel selection, which consists of several different functional units, measurements and recordings of the error rate along shortwave paths of different length are required. By comparing a known transmitted text (e.g. 1022 bits per test period) with the demodulated message, the error rate at the receiving end can be determined in each of the three channels. Furthermore it is at the same time necessary to record the error rate after channel selection and error correction in the error corrector so that the circuitry that helped reduce the error rate can be recognized. Figure 105 shows the test results at five test points on a radio link. This illustrates – from the top working down – the error rate in channels 1, 2 and 3, the error rate after channel selection and the residual errors after error correction. A transmission time of one hour has been plotted and the transmission rate on each channel was 100 bit/s. It can be seen that during this time channels 1 and 2 were equally affected and had a considerably lower error rate than channel 3. The error rate of the selected channel is a great deal lower than the error rate of the three separate channels. After a transmission time of approx. 10 minutes, broadband disturbances occur which cause several errors after channel selection. The forward error

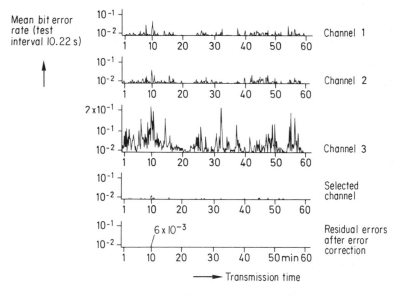

Fig. 105   Mean bit error rate as a function of the transmission time

147

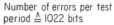

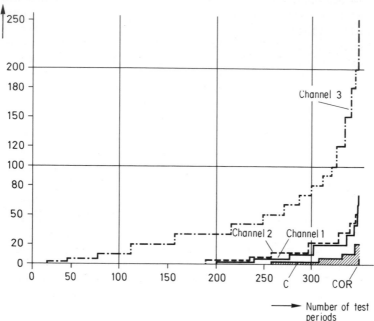

Fig. 106   Number of errors per test period measured over a long time interval

correcting device is overtaxed by a disturbance which lasts too long, which in this case leads to a total of three residual errors.

Figure 106 shows another interpretation of the measurements over the same shortwave radio link. It can be clearly seen that channels 1 and 2 are roughly equally disturbed whilst the third channel is affected by considerably more errors. The relative bit error rate of the selected channel is an improvement by a factor of 15 on the mean relative bit error rate of approx. 1.4×10^{-2} on the three channels. After error correction with the aid of the convolutional code, a relative residual error rate of 1.6×10^{-5} results, the factor for improvement on the mean bit error rate in the three channels being 835.

7.8   ARQ Protection Methods for Simplex Operation

ARQ methods requiring a duplex connection have been described in Section 7.5. Message exchange is possible here in both directions

148

simultaneously. To repeat signals falsified *en route* a backward channel must always be available for transmitting the RQ signal.

Where trouble-free duplex operation cannot be ensured, e.g. because antennae are located too close together* or because of a lack of radio frequencies, an ARQ simplex method may be employed to protect the transmission circuit. This method permits one-way-at-a-time transmission, however, with a backward path being provided for the transmission of acknowledgment signals. To this end the message flow must be periodically interrupted at the transmitting station so that acknowledgment signals can be received from the distant station. The message to be transmitted is divided into blocks of, for instance, three characters. Following each block, acknowledgment signals are sent in the opposite direction which report correct or incorrect transmission and effect channel turn-around. The repetition cycle of an ARQ simplex system is shown in Fig. 107. To obtain a message flow of e.g. 50 bauds between terminal units, the transmission speed on the radio link must be increased to about 100 bauds, taking into account the signal delay on the transmission circuit and the inherent delay of the radio equipment. Periodic turn-around requires the use of radio equipment, transceivers† for instance, which can be alternately switched to transmission and reception within a few milliseconds.

Figure 108 shows a shortwave radio link for simplex traffic with ARQ data protection.

For error detection the ARQ simplex systems employ a constant-ratio 7-unit code, the start and stop polarity elements occurring at a ratio of 3:4.

7.8.1 Operating Principle of an ARQ Simplex Terminal

Two stations, A and B (Fig. 107), interwork via a shortwave radio circuit in the synchronous mode so that at a particular time one of them is operating as the message sending station and the other as the message receiving station. Figure 109 shows what an ARQ simplex terminal looks like. As soon as the connection has been established in one direction, the message to be transmitted is split up into blocks of four characters each, for example. These blocks are marked differently, i.e. they are transmitted alternately inverted and not inverted so that the receiving station can be synchronized with the sending station. At the start of a transmission, the characters A, B,

---

* Antennae operated in close proximity to each other occur frequently in mobile service and on board ships.
† Transceiver: shortwave transmit/receive equipment for simplex traffic.

150

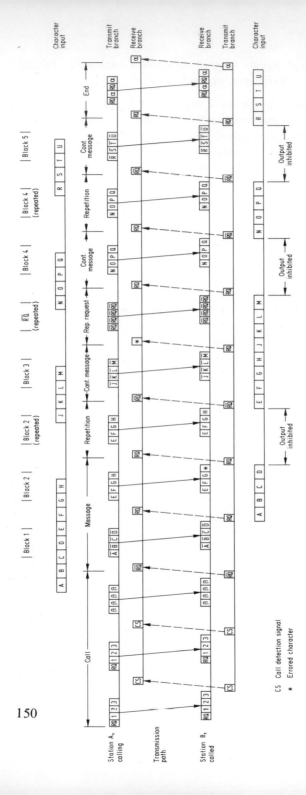

Fig. 107 Repetition procedure for ARQ simplex operation

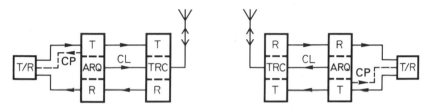

Fig. 108 Shortwave radio link for simplex traffic with ARQ data protection

ARQ   ARQ simplex system
T, R   Transmit/receive facilities
TRC   Transceiver
T/R   Transmit/receive teleprinter
CL   Control line for radio channel turn-around
CP   Call-down pulse

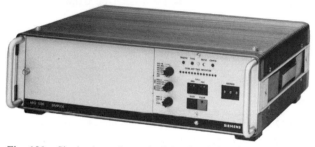

Fig. 109 Single-channel terminal ARQ 1000 S for simplex operation

C and D, for example, are marked as block 1 (inverted). The receiving station B checks the 3:4 ratio of each transmitted character. If the characters were not falsified when transmitted, the receiving station forwards them to the customer terminals. At the same time the transmit branch of station B returns an acknowledgement signal RQ (not inverted) to station A to confirm that block 1 has been received error-free and that transmission of block 2 can be commenced. Figure 107 shows the procedure involved in transmitting from station A to station B, as well as repetitions in the event of a falsified character being received.

Station A now checks the 3:4 ratio of the acknowledgement signal RQ. If this signal has been received error-free, the next signals, for example E, F, G and H, can be transmitted as block 2 (non-inverted).

If in the event of an errored transmission, e.g. in the second block, character H is falsified, block 2 is not forwarded to the customer's terminal

151

equipment. Instead, station B sends the repetition request signal RQ in place of the acknowledgement signal $\overline{RQ}$ as an invitation to transmit block 2 again. If the signals E, F, G and H are then received without error they are forwarded to the terminal equipment. The acknowledgement signal $\overline{RQ}$ is transmitted to station A as confirmation of correct reception. In the continuing transmission, signals J, K, L and M are sent in inverted form as block 3 and, if they are received correctly, are acknowledged with RQ (non-inverted).

If during a transmission the acknowledgement signal RQ, for example, is falsified, four RQ signals are sent to station B (as shown in Fig. 107) to invite station B to transmit the acknowledgement signal once more.

Should the signals in a block **and** the subsequent acknowledgement signal all be falsified, station A continues to transmit RQ signals until the appropriate acknowledgement signal is received error-free from station B.

While a transmission is in progress, both station A and station B can initiate the OVER procedure to enforce a reversal of the direction of transmission (Fig. 110).

ARQ transmit branch

The operation of the ARQ simplex terminal will be explained with the aid of the block diagram in Fig. 111. The message to be transmitted is sent by the data terminal equipment in the same rhythm as the pickup pulses in the form of characters transmitted in series and applied to input circuit IN. Calldown circuit COU releases the characters if the two stations are in synchronism and are not cycling. The characters transmitted by the data terminal are first of all converted into parallel characters by series–parallel converter SPC. Character storage CS, which is next in the circuit, has a capacity of 1024 characters. When this is filled to 95% of its capacity, a lamp lights up which warns against further data input and thus prevents storage overflow. A second lamp signals 'storage full' and stops the emission of pickup pulses to the sending machine. If the characters are input from a data source not designed for controlled signal pickup, this storage acts as a buffer which permits continuous data flow between customer and ARQ equipment. The message is subsequently transferred from this buffer storage to the repetition storage RS. The latter assembles the character blocks required for ARQ operation and hands them over to the coder COD. At the same time these blocks are kept in storage ready for repeat transmission should this be necessary. Continuous start polarity at

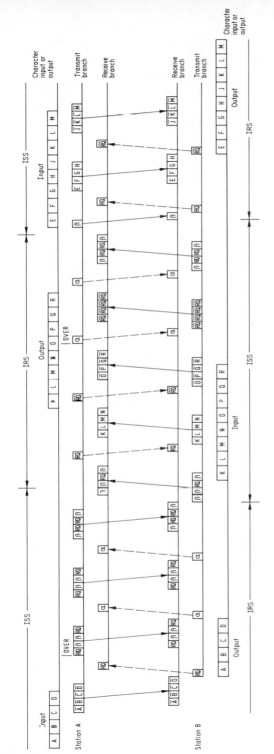

Fig. 110 Circuit turn-around with an ARQ simplex connection

OVER Circuit turn-around
ISS   Information sending station
IRS   Information receiving station

153

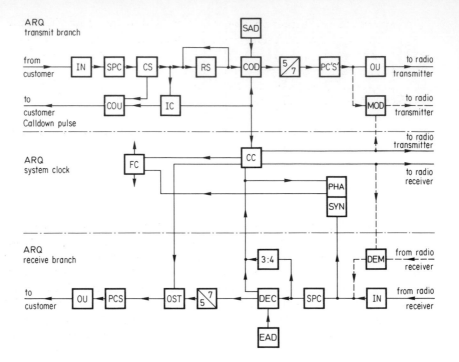

Fig. 111   Block diagram of an ARQ simplex terminal

ARQ Transmit Branch		ARQ Receive Branch	
IN	Input circuit	IN	Input circuit
SPC	Series–parallel converter	DEM	Demodulator
CS	Character storage	SPC	Series–parallel converter
COU	Calldown pulse output circuit	DEC	Decoder
IC	Input control	EAD	Receive address
RS	Repetition storage	3:4	Code check
COD	Coder	7/5	7/5 code translator
SAD	Send address	OST	Output storage
5/7	5/7 code translator	PCS	Parallel–series converter
PCS	Parallel–series converter	AS	Output circuit
OU	Output circuit		
MOD	modulator		

ARQ System Clock

CC	Common control
PHA	Phasing circuit
SYN	Synchronizing circuit
FC	Frequency clock

154

the input of the ARQ transmit branch is translated by the coder COD into the idle signal α, and continuous stop polarity into idle signal β. In addition, the coder forms the control signal RQ for the repetition request, as well as the call and clearing signal combinations and the turn-around control signal (OVER). The 5-unit code signals are subsequently translated in the code translator 5/7 into 7-unit code signals at a ratio of 3 stop:4 start polarity elements. The parallel–series converter PCS converts the parallel signals into the serial signals required for transmission. In the output circuit OU the message is conditioned for dc keying or, in modulator MOD, for single-sideband or F1 operation, and passed on to the radio transmitter.

ARQ receive branch

The 7-unit code signals arriving from the radio receiver are fed to the input circuit IN of the ARQ receive branch. A single-sideband or F1 message, on the other hand, is first translated into dc signals by demodulator DEM. The phase relation of the character elements and of the characters is supervised in synchronizing circuit SYN and in phasing circuit PHA and, if necessary, corrected. After the conversion of the serial signals into parallel ones in series–parallel converter SPC a code check is performed. Any polarity ratio other than 3:4 leads to a repetition process which is directed by common control CC. At the same time the message is passed on to decoder DEC.

The decoder recognizes the incoming call and compares it with the receive address EAD specified. If the two are identical, the called station is activated and synchronizes itself to the calling station. The decoder also interprets the control signals required for operation. Reception of the idle signals α or β initiates the transmission of steady start or steady stop polarity to the customer. Upon detection of the query signal RQ the common control starts a repetition process. Behind the decoder the 7-unit code signals are reconverted, in the 7/5 code translator, into their original 5-unit code configuration required for operating the teleprinter. The succeeding output storage OST supplies the message signals only as long as no repetition cycle takes place. In parallel–series converter PCS the parallel signals are once more converted into serial ones and, after the start and stop elements have been added, are sent to the customer via output circuit OU.

ARQ system clock

The system clock supplies the control pulses for the ARQ or FEC system. A crystal oscillator operates at a frequency of 3686.4 kHz with a maximum drift of 2×10^{-6}. For the desired transmission rate, adjustable to 50 or 100 bauds, clock pulses for the send and receive branches are generated in the frequency clock FC, and these control the sequence of operations in the single-channel terminal. In the slave station of an ARQ connection the clock pulses for the send leg and the receive leg are derived in common from the receive-side frequency divider. A phasing circuit PHA and a synchronizing circuit SYN also belong to the system clock. These ensure that the communicating stations operate synchronously.

FEC operation

In the ARQ simplex system described above it is also possible to employ forward error correction (FEC) for protecting telegraph messages and data on one-way shortwave connections or in broadcast services. Messages are transmitted on the time-diversity basis (Fig. 112). The initial transmission of a character is defined as **direct transmission** and the second transmission as **repeat transmission.** Repeated characters have their polarities inverted with respect to those directly transmitted. The spacing between direct and repeat transmissions amounts to 15 characters. This means that at a customer rate of 50 bauds – transmission rate of 100 bauds – all characters can be corrected in the event of noise bursts lasting up to about 1 s. Code combination SPACE is transmitted to the customer when an errored character is detected.

Connection setup

A calling facility (Fig. 107, station A) permits any station B with a permanently assigned (preset) address which is in the standby condition to be switched on or off, and thus allows a connection to be set up automatically. In the broadcast mode (FEC) all subscribers may be called simultaneously. The call address consists of three digits and an address framing signal RQ (Figs 107 and 112). As a protection against transmission errors, the call address must be received several times in succession (at least twice) before the equipment can interpret the call as recognized. Connection setup is initiated by depressing the ARQ call button. The radio transmitter of the calling station is automatically cut in under the control of the ARQ terminal. In the ARQ mode, the called station returns an acknowledgement signal CS after receiving the call. If a call does not result in a connection being established, the calling station automatically returns to the ready-to-receive condition.

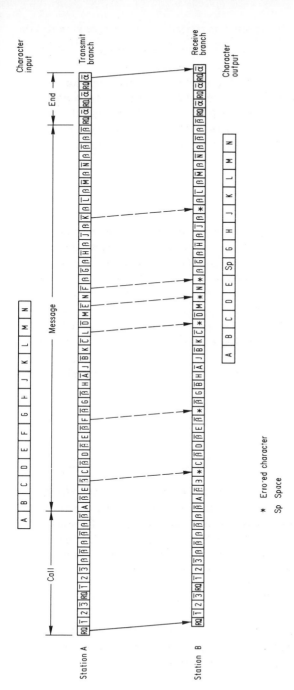

Fig. 112 Diagrammatic representation of FEC operation

* Erro'ed character

Sp Space

157

In the FEC mode, a call is interpreted as recognized in the receiving station when the address has been correctly received twice within a certain period of time.

An existing connection is cleared down by transmitting the end-of-operation code (Figs 107 and 112). In the ARQ mode, cleardown may be effected from either end. In the FEC mode, on the other hand, this is only possible from the sending station. At the end of this cleardown procedure all stations return to the standby condition and the radio transmitters are deactivated.

Input control

With the input control EST (Fig. 111) the ARQ terminal can be operated in a dialogue with the teleprinter. This is particularly advantageous for distant customers. The customer is then able to select the operating mode and to input operation control commands as well as the subscriber address, using a teleprinter. The subscriber address consists of a three-digit number. The connection can be set up either in the ARQ or FEC mode, the operating mode being selected by inputting the letter sequence ARQ or FEC. The OVER signal combination is provided to reverse the direction of transmission. After transmission has been completed, the connection is cleared by inputting the signal combination CLEAR. At the start of transmission the teleprinter logs the selected operating mode (ARQ or FEC), the subscriber address, and whether or not the connection was set up.

7.9 Comparison Between the ARQ and FEC Protection Methods

The ARQ and FEC protection methods depend on different redundant codes for error detection and error correction, which permit the character error rate to be reduced on disturbed transmission paths. The degree of improvement in the quality of a transmission is a function of the correction power of a given protection system and of the grade of service and character error rate of the shortwave circuit involved.

In forward error correcting systems in particular the distribution in time of the bit errors must not be neglected. In data transmission systems employing automatic repetition request (ARQ systems) consideration must be given not only to the character error rate but also to the transmission efficiency, a decrease of which causes the data transfer rate to be reduced.

To achieve a mean efficiency of 80% in an ARQ-protected shortwave radio circuit and a character error rate of less than 10^{-5}, the mean quality of the transmission channel must not drop below a certain value (see Chapter 7.5.4). Forward error correcting systems likewise require a certain minimum quality of the transmission channel, e.g. $P \leq 10^{-2}$. If the transmission path of an ARQ-protected connection is affected by excessive interference, message flow may be completely blocked due to continuous cycling even when the character error rate is as low as 10^{-5} to 10^{-6}. In this case the efficiency comes down to zero. With the FEC method, on the other hand, the information flow remains intact. The character error rate, however, depends on the transmission path. In the event of abnormal interference, the error correcting circuit may be overtaxed and the signals passed on to the terminal equipment may be falsified. An automatic disconnect facility will then suspend onward relaying of the message to the terminal equipment and resume this function only when the transmission quality of the connection has again reached a certain level. The efficacy of an FEC system has been described in Chapters 7.6.1 and 7.7.1.

7.10 Data Terminal Equipment for ARQ Systems

The terminal devices for data input and output used on unprotected shortwave radio links are teleprinters or perforated tape devices in most cases, which are connected direct to the transmission facilities. Where teleprinter messages cannot be immediately fed to the transmission equipment, as in the case of all ARQ-protected shortwave radio links, terminal devices suitable for call-down under step-by-step control must be used unless storage facilities are interposed.

The use of ARQ data protection equipment requires buffering of the message at the transmitting end. In this way the message flow may be interrupted at any time, e.g. for phasing the ARQ terminals over the radio link or for repeating errored signals. This buffering requirement can be met by call-down terminal devices or by electronic interim storages which must be designed for individual signal pick-up.

Apart from the recently introduced electronic storages, such as telegraph storages holding 4000 characters, electromechanical reperforator/transmitters are still in use. The latter work with 5-unit code tape. The perforated tape, serving as storage medium, is accommodated in a cassette. For convenient reading, a clear-text monitoring print is provided on the tape. The reperforator/transmitter permits the message stored in the tape to be transmitted in response to command signals, independent of the incoming message flow.

The CCIR recommends a storing capacity of 750 characters for storages used in connection with ARQ terminals.

The storages are designed to provide status indications such as empty, full and nearly full.

Tape transmitters designed for call-down operation are preferably used where terminal devices are to be connected immediately to the ARQ equipment. The messages in the previously punched tape are then called down from this unit by the ARQ terminal emitting a control signal for each individual character.

Teleprinters (e.g. Teleprinter 1000) suitable for call-down operation in connection with ARQ terminals will be used for keyboard and tape transmissions.

7.10.1 Telegraph Storage

The telegraph storage has the task of buffering messages when these are to be forwarded regardless of their time of arrival. This is always the case with ARQ systems if the system is performing a repetition cycle because errored characters were received. Figure 113 shows the buffer storage as a send storage. The messages can be called down character-by-character. Besides its buffering task, the telegraph storage can also be employed as a speed converter (Fig. 114). The efficiency of message transmission can be enhanced – from the customer's point of view – by increasing its speed. In the error-free transmission periods the message can thus be transmitted at a higher speed, allowing the repetition times of the ARQ system to be compensated for. The first storage unit ST 1 then acts as a send storage and

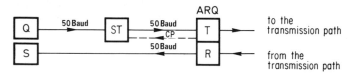

Fig. 113   Use of the telegraph storage as a buffer storage on ARQ dedicated circuits

Q	Source	T	Transmit branch
S	Sink	R	Receive branch
ST	Telegraph storage	CP	Calldown pulse
ARQ	Data protection system		

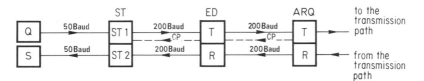

Fig. 114   Use of the telegraph storage as a speed converter on ARQ dedicated circuits in crypto operation

Q	Source	ARQ	Data protection system
S	Sink	T	Transmit branch
ST	Telegraph storage	R	Receive branch
ST 1, 2	Storage units 1, 2	CP	Calldown pulse
ED	Encryption device		

Fig. 115   Telegraph storage FSP 4000/2 in carrying case

the second one ST2 as a receive storage, i.e. the latter reconverts the speed. Figure 114 shows message transmission at a customer speed of 50 bauds, while the speed on the transmission path is 200 bauds. It also shows the point on the transmission path at which any encryption device ED which may be necessary must be inserted. Figure 115 is a picture of the telegraph storage FSP 4000/2, which has two storage units each with a capacity of 4096 5-unit code telegraph characters.

Principle of operation of a telegraph storage

The functional method of a telegraph storage is illustrated, using a block diagram (Fig. 116). The messages input into storage unit 1 in serial form are fed via the input circuit IN to series–parallel converter SPC, and from there to the storage ST. The storage is addressed through control circuit

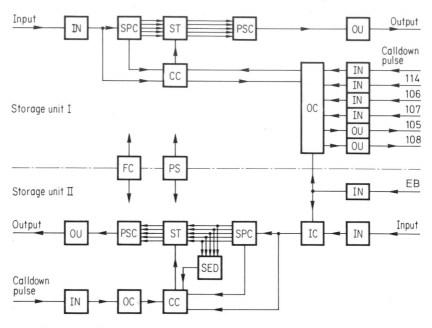

Fig. 116   Block diagram of a telegraph storage

FC	Frequency clock	PSC	Parallel–series converter
PS	Power supply	OU	Output circuit
IN	Input circuit	OC	Output control
SPC	Series–parallel converter	SED	Start/end detector
ST	Storage	IC	Input control
CC	Control circuit		

CC which prevents the continuous start and continuous stop polarity conditions being entered into it. If required, continuous start polarity can be switched through to the output when the storage is empty. The messages are put out serially and uninterruptedly with normal or 50%-prolonged stop elements by the output control OC via parallel–series converter PSC and output circuit OU. With ARQ operation pickup can be effected character-by-character or using a modem for transmission over a line, in accordance with the interface definitions as per V.24 (interchange circuits 114, 106, 107, 105, 108, EB).

Storage unit 2 is structured like storage unit 1. In ARQ operation it can be employed as both a send and a receive storage, whilst in the case of operation using a modem it always acts as a receive storage. Some of the

162

V.24 signalling can consequently be dispensed with in storage unit 2. The start/end detector SED in storage unit 2 makes it possible to commence or terminate entering data into the storage with the aid of an address.

In order to be able to see how full both storage units are at any time, the following storage conditions are indicated: 'storage seized', 'early warning' and 'storage full'. Each storage unit is provided with a separate delete button which is guarded to prevent unintentional deletion.

7.10.2 Teleprinter 1000

The electronic Teleprinter 1000 is a page-printing terminal which complies with international conventions on teleprinter traffic. The characters are coded in accordance with ITA code No. 2. It has been designed for operation both in the Telex networks and in radio telegraph networks (Fig. 80). In connection with error control systems (see Chapter 7.4) the Teleprinter 1000 may be used for signal transmission under the control of calldown pulses.

The Teleprinter 1000 can be adapted to telegraph speeds of 50, 75 and 100 bauds. In off-line operation the speed is invariably 100 bauds and the machine is here used for the preparation of perforated tape intended for subsequent transmission over the line.

The line terminating circuitry for the different interface options and applications is integrated in the machine. Operational functions are controlled by a microprocessor. A relay unit with a telegraph socket may be provided in addition. In connection with the send/receive switch in the line control module of the Teleprinter 1000, a contact may be brought out which, constituting the interface to external devices, may be used for the on/off control of radio transmitters or for switching over transceivers from transmission to reception.

8 Test Methods for Determining the Residual Error Rate

To be able to compare the residual error rate of the various transmission methods, measurements must be carried out on different shortwave radio links over different distances and with different frequencies. To improve the transmission quality on shortwave radio links, diversity and data protection systems may be employed singly or in combination.

Since the transmission conditions on shortwave radio links vary greatly with time, the error rates of the methods to be compared must be recorded simultaneously to permit an accurate assessment of the improvement capabilities inherent in the different methods. Two measuring options are indicated.

For contrasting measurements with two transmission methods, a multi-channel VFT terminal for radio operation (WTK) may be employed which affords two transmission channels of matched bandwidth. The two channels of **equal bandwidth** should be contiguous in the frequency assignment so that they will be affected to the same degree, or nearly so, by interference phenomena. The result of this comparison may, however, be falsified by selective interferers hitting one channel more than the other. It should be noted that the output power of the radio transmitter is split up and therefore lower for each channel. The test signals employed may be 5-unit code

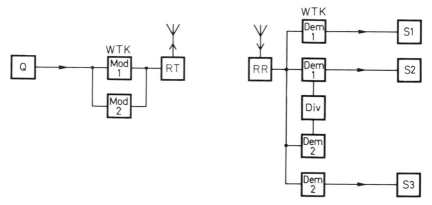

(a) Comparison between single-channel and frequency diversity reception (class of emission A7J)

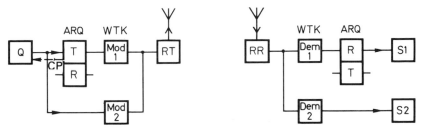

(b) Determining the error rate of unprotected transmissions and of ARQ-protected transmissions. Schematic representation of an ARQ duplex circuit (class of emission A7J)

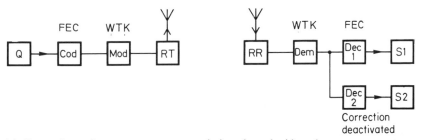

(c) Comparison of error rates on a transmission channel with and without error correction

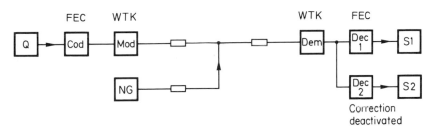

(d) Determining the error rate in the presence of white noise with and without error correction

Fig. 117   Test set-up for determining the error rate of the different transmission methods

Q	Source	Div	Diversity combining circuit
S1, 2, 3	Sink	ARQ	ARQ terminal
WTK	VFT transmission terminal	T/R	Transmit/receive facilities equipment
Mod 1, 2	Modulator	FEC	Forward error correction
Dem 1, 2	Demodulator	Cod	Coder
RT	Radio transmitter	Dec 1, 2	Decoder
RR	Radio receiver	NG	Noise generator
ET	External tripping	CP	Calldown pulse

teleprinter messages of a defined text and of known length (e.g. test patterns comprising 1000 characters) or signals from a data test set having a defined number of bits per measuring cycle (e.g. 1022 bits). Both channels should be loaded with the same test pattern.

To eliminate the disadvantages of the above mentioned contrasting measurement, the shortwave radio link may be loaded with a single transmission channel. This makes it necessary to use two demodulators and two data protection units at the receiving end. The two messages received in this way are then subject to exactly the same source and degree of interference.

To ascertain the behaviour of shortwave transmission facilities in the presence of disturbing signals without an actual radio circuit being interposed (Fig. 87d) or to compare the different data protection methods, 'white noise' may be employed as a standardized noise source. White noise comprises a large number of frequencies in the range between near to zero Hz and values whose upper limit is determined by the circuits to be measured and by their components. For measuring the error rate as a function of the normalized signal-to-noise ratio, noise of a certain power rating is applied to the input of the WTK demodulator in addition to the useful signal of defined magnitude.

Contrasting measurements of the transmission characteristics employing white noise only apply in analogy to the described measuring methods depicted in Fig. 117(a)–(c).

To obtain meaningful results from these contrasting error-rate measurements on shortwave radio links, the number of the characters or bits to be transmitted must be such that at least 1000 errors will occur in the course of these transmissions. Examples are given in Fig. 117(a)–(d).

166

9 Radio Telegraph Networks

Fixed and mobile shortwave communication networks are finding increasing application, e.g. for special services. They have a radial configuration in most cases. The individual stations may be equipped with data protection devices and operate on a duplex, simplex or one-way basis. Since the number of the frequencies available in the shortwave range is limited, voice frequency telegraph transmission systems are preferably used (Chapter 5). These systems permit a number of channels to be transmitted simultaneously over **one** assigned radio frequency. Special transmission equipment (WTK) has been developed for operation on shortwave radio links to account for the propagation conditions in this frequency range.

The example given in Fig. 118 shows a communication centre from which a number of independent messages are transmitted by a single radio transmitter to several out-stations in a radial configuration. Single-sideband modulation with suppressed carrier (A7J) is employed for transmitting the

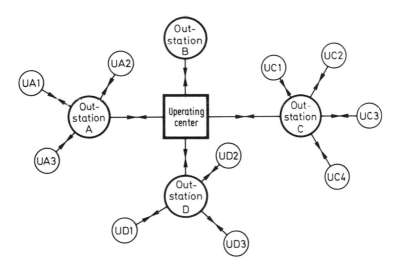

Fig. 118   Example of a radial shortwave radio network comprising a communication centre, four out-stations and a number of sub-stations (UA1–UD3)

167

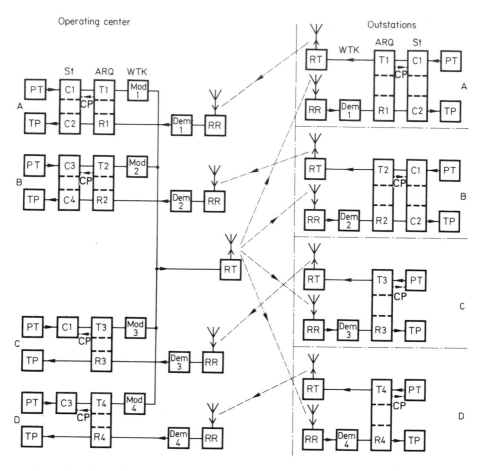

Fig. 119 Block diagram of a radial network comprising a communication centre and four out-stations

A, B, C, D	Radio telegraph links	T/R 1–4	Transmit/receive facilities
PT	Punch tape equipment (with or without call-down facility)	WTK	VFT terminal for radio operation
TP	Teleprinter	Mod 1–4	Modulators
C 1–4	Channels of storage	Dem 1–4	Demodulators
St	Storage	RR	Radio receiver
ARQ	ARQ terminal	RT	Radio transmitter
		CP	Calldown pulse

messages. Depending on the transmission speed, up to four channels (with a frequency shift of $\pm$ 85 Hz per channel) can be accommodated in a voice band. The frequency gap between the channels should be greater than 500 Hz to avoid the risk of intermodulation. As regards the transmitting power, it should be noted that the power available for each channel decreases with the square of the number of the channels.

At the receiving stations the messages are received as quasi-F1 transmissions selectively by means of a radio receiver and an F1 demodulator. The out-stations depicted in Fig. 118 in turn function as sub-centres for their associated sub-stations.

Figure 119 shows the block diagram of a radial shortwave radio network with the facilities in the communication centre, with the operator's positions A through D, and the four out-stations A through D (see Fig. 120).

Fig. 120 Receiving station in a communication centre for the reception of protected data transmissions

ARQ terminals are employed to protect the messages on the individual radio links. The enquiry procedure requires two stations to be permanently assigned. The number of the transmitters and of the frequencies required may be reduced by a VFT terminal for radio operation (WTK) to be installed in the centre. The messages for the out-stations are each assigned to one WTK channel and radiated in the A7J mode. The out-stations are equipped with radio receivers whose output is fed to an F1 demodulator as previously mentioned. Only the WTK channel actually intended for it is picked out of the channels transmitted by the respective station and the message is received as an F1 transmission. The return legs from the out-stations to the centre operate with class of emission F1.

The terminal devices used in the centre and in the out-stations for data input and output are teleprinters and perforated tape equipment designed for call-down operation. Crypto equipment may be interposed to prevent malicious external listening.

In place of the call-down teleprinters, equipment without a call-down facility may be employed if a buffer storage for individual signal pick-up, a telegraph storage for instance, is provided. The storing capacity of the storage shown in Fig. 119 amounts to 4000 characters per communication channel.

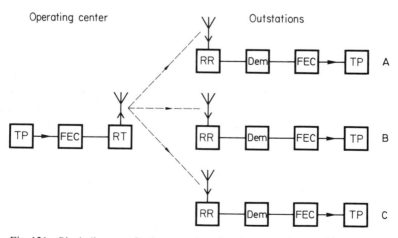

Fig. 121   Block diagram of a shortwave broadcasting network comprising one communication centre and three out-stations

Fig. 122 Radio telegraph transmission terminal for telegraphy and data transmission. From top to bottom: telegraph storage FSP 4000/2, data protection system ARQ 1000D, and VF terminal WTK 1000

In the recent past 'high-speed message transmission' has been introduced on shortwave radio links. The message originally delivered at a telegraph speed of 50 bauds is here accepted into a storage for subsequent retransmission at an increased speed (100 bauds or 200 bauds, for instance). The storage is thus used as a 'speed converter' on radio links A and B (Figs 119 and 122). At the receiving end the message is again loaded into a storage where it is converted back from e.g. 200 bauds to 50 bauds before it is passed on to the terminal station. Transmission gaps caused by excessive interference or fading on the radio path can in this way be bridged and this means practically an improvement in the efficiency of the radio link so far as the out-stations are concerned.

In broadcasting networks (for meteorological or news services) messages are mostly transmitted from a central location. In this case forward error

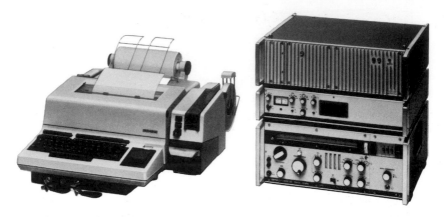

Fig. 123   Radio receiver station for the reception of protected F1-transmission with the Teleprinter 1000 CA

correcting equipment (Section 7.6) lends itself particularly well to data protection because a backward channel for transmitting acknowledgments is not required. Figure 121 shows the facilities provided for a communication centre and three out-stations (one out-station is shown in Fig. 123). It is possible to call these stations either simultaneously or in groups or, employing a digital calling system, each station may be selected individually.

10 Examples of Radio Telegraph Links

Figures 124–130 show examples of shortwave radio telegraph links. Terminal equipment and transmission facilities have been discussed in detail in the preceding chapters so that no further elaboration is required.

Single-channel connections to a distant station both without and with data protection are shown in Figs. 124–126. Figure 127 depicts an error-corrected multichannel circuit operating on a time-division basis by means of ARQ terminals while a one-way radio circuit with forward error correction is shown in Fig. 128. The simultaneous transmission of a number

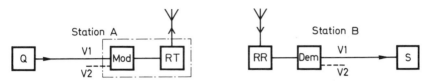

Fig. 124 Unprotected telegraph transmission to a single station (point-to-point connection) employing the F1, F4 and F6 classes of emission

Figs 124–130 show:

Q	Source	Div	Diversity combining equipment
S	Sink	St	Storage
Mod 1–3	Modulator	CP	Calldown pulse
Dem 1–3	Demodulator	ARQ	ARQ terminal
RT	Radio transmitter	A, B; V1, V2	Communication channels
RR	Radio receiver	FEC	Forward error correction
AD	Antenna diversity	T, R	Transmit/receive facilities

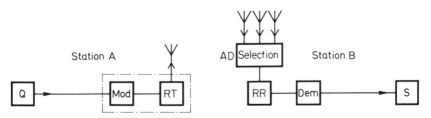

Fig. 125 F1 reception, antenna diversity

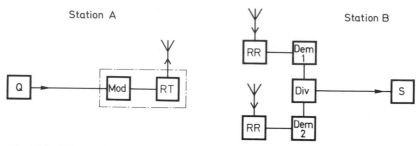

Fig. 126   F1 reception, space diversity

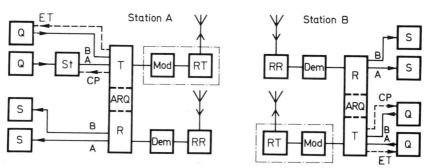

Fig. 127   F1 reception from one co-station with ARQ protection (duplex circuit)

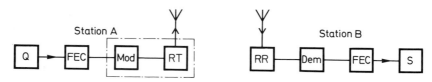

Fig. 128   One-way connection, F1 class of emission with FEC data protection

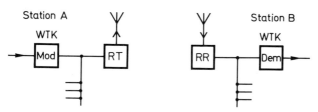

Fig. 129   Multichannel connection with WTK channels in single-sideband operation

174

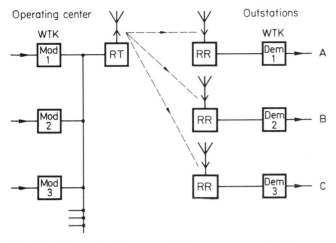

Fig. 130   Communication centre with three out-stations (radial network)

of telegraph channels within a voice band in single-sideband operation on a frequency multiplex basis is shown in Fig. 129. The data-protection methods may be employed here as in the preceding examples.

Figure 130 shows a radio network which permits simultaneous transmission to several out-stations. Class of emission A7J and a cor-

Table 12   Classes of emission, channels, frequency shift and telegraph speeds

Class of emission	No. of channels	Frequency shift	Telegraph speed	Remarks
F1	1	± 50 Hz to ± 200Hz	50 to 200 bauds	
F4	1	± 400 Hz	max. 3600 bauds	Facsimile
F6	2	± 50 Hz, ± 150 Hz to ± 200 Hz, ± 600 Hz	50 to 100 bauds	
A7A, A7J WTK channel FM 170	Max. 16	± 42.5 Hz	Start-stop 50 to 75 bauds	Synchronous 50 to 100 bauds
A7A, A7J WTK channel FM 340	Max. 8	± 85 Hz	Start-stop 100 to 150 bauds	Synchronous 50 to 200 bauds

responding number of VFT channels are used. Other variants are also possible for the above examples.

Table 12 compiles the channels afforded by the various classes of emission as well as the frequency swings and modulation rates usually employed.

10.1 Frequencies for Shortwave Links

Selection of the proper **operating frequencies** is a decisive factor in planning shortwave radio links. The incoming signal level is determined, apart from the distance between the inter-operating stations, by the send and receive antennae used and the propagation conditions applying to the chosen operating frequencies. The propagation conditions vary in the course of a sunspot period and they vary also with the time of the day and the year.

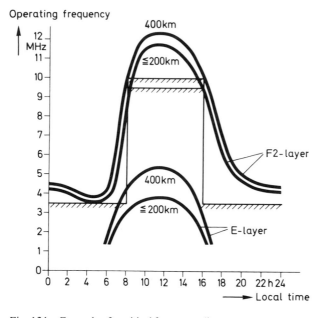

Fig. 131   Example of a critical frequency diagram

Location	50° North, 10° East
Time of year	Winter
Sunspot number R_{12} (max.)	150
/////////	Chosen operating frequency

176

The operating frequencies to be used with sky-wave propagation over a period of several hours are selected on the basis of the **critical frequency diagrams** applicable to the transmission route. These diagrams are derived from the **radio propagation predictions.** Figure 131 gives an example of a critical frequency diagram for radio links covering a distance of from 200 to 400 km. For 24-hour operation the frequencies within the range from 3.5 MHz to 10 MHz, for instance, may be used as operating frequencies, resorting to changeover at 8 h and 16 h.

The received field strength obtained in specific cases at the receiving station is shown in Fig. 132 for the operating frequencies 3.5 MHz and 10 MHz. At the highest point of the sun the ionization of the reflecting layers is greatest and, consequently, the attenuation is reaching its peak. The incoming signal level is therefore at a minimum during the hours around noon. Fading may cause a decrease in the field strength by about 20 to 40 dB. Considering the

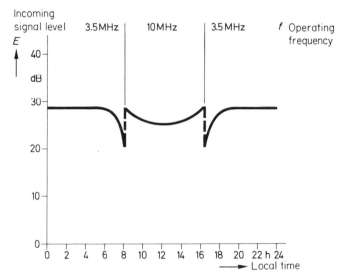

Fig. 132   Reception field strengths for a 24-hour shortwave radio connection with the same cut-off frequency as that in Fig. 110

Location	50° North, 10° East
Time of year	Winter
Sunspot number R_{12} (max.)	150
Length of radio link	400 km
Transmitting power	100 W
Antenna gain	$G_D = +6\,dB$

177

decrease in field strength during the summer months (in particular during the hours around noon), a signal adequate for proper reception at any time can no longer be ensured with a limited send power. As can be seen from Fig. 131, 24-hour radio operation requires a changeover from day frequency to night frequency.

10.2 Transmission of Encrypted Messages

Communication in all areas of public life opens up the possibility that transmitted information can be misused. In both the military and diplomatic sectors, important information has been transferred in encrypted form for centuries. The most effective method of making information incomprehensible to unauthorized people is encryption. Encryption equipment developed for this purpose permits information to be transmitted from sender to recipient without the danger of it being intercepted. Today there is encryption equipment for the wire-bound transmission of telegraph characters, data and facsimile messages, as well as for digitized voice communication. This equipment can also be used for shortwave transmission. As Fig. 133 shows, the encryption equipment is then interposed between the source and the transmission facilities.

To transmit a cryptogram, the information is converted on the send side into a form which is unrecognizable for an unauthorized person. At the distant station the message must then be reconverted into the original text. This is only possible if the encryption equipment on the send side is in step with the encryption equipment on the receive side. This synchronism of the send and receive sides is known as crypto synchronism which can be disturbed if characters are lost or added during transmission. If this occurs, decrypting of the complete transmitted message on the receive side is difficult or no longer possible.

The transmission of encrypted messages over shortwave radio links

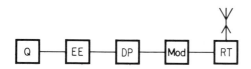

Fig. 133   Linking up encryption equipment in the transmission path (send side)

Q	Source	Mod	Modulator
EE	Encryption equipment	RT	Radio transmitter
DP	Data protection equipment		

178

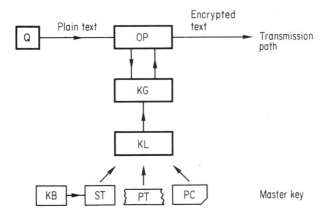

Fig. 134 Block diagram of an encryption device

Q	Source	KB	Keyboard (e.g. Teleprinter
OP	Input/output and operating		Model 1000 CA)
	procedure	ST	Storage
KG	Crypto generator	PT	Paper tape
KL	Key loader	PC	Punched card

therefore calls for particularly high transmission reliability. This can be achieved by employing data protection equipment (Chapter 7).

Figure 134 shows a block diagram of encryption equipment. The data source supplies the original text which is to be sent, i.e. the plain text, to the input circuit of the encryption device in digital form as a telex, facsimile or other data. The crypto generator KG encrypts and decrypts the information. It must be reliable from a cryptological point of view. Using a certain algorithm, the encrypted text is produced as a quasi-random bit sequence with a great periodicity in order to achieve a high degree of crypto security. In the encrypted text it must not be possible to detect what rules have been followed.

In order to set the encryption equipment at both the send and receive stations to the same key, a message key must be transmitted at the start of operation. This is formed in the send station from a specific number of random characters. Each customer also enters his own daily keying variable (master key) by way of the key loader attached to the equipment. Depending on the type of equipment, this key can be input using paper tape, punched cards, magnetic cards or the keyboard of a teleprinter (e.g. Teleprinter Model 1000 CA, Fig. 135). The daily key must be protected

Fig. 135 Teleprinter Model 1000 CA with encryption device (in base tray) for cryptological applications

from unauthorized access. The encrypted text sent to the transmission path is produced from the plain text input and a quasi-random bit sequence, generated in the crypto generator, and determined by the algorithm and the keysetting (daily key and message key). Encryption devices are generally employed as independent devices on a transmission path. Figure 135 shows a Teleprinter Model 1000 CA for cryptological applications which operates fully electronically and automatically and into which the encryption device has been integrated.

Different encryption equipment is used for the transmission of different types of message. Table 13 shows the range of speeds at which encrypted messages can be transmitted.

Table 13   Baud and bit rates and type of encrypted message

Type of message	Typical transmission speeds
Telex	50, 75, 100 bauds
Computer data	50 to 300 bauds, 600, 1200, 2400, 4800 and 9600 bit/s
Facsimile	2400 bit/s
Digitized speech	2400 bit/s

180

In order to be able to transmit speech information reliably from a cryptological point of view, the voice signal must be digitized prior to encryption and then transmitted in this form. For this purpose vocoders have been developed which employ an analysing and synthesizing process. This process permits the relatively low transmission speed of 2400 bit/s and, in conjunction with appropriate shortwave transmission facilities, allows encrypted speech to be transmitted over shortwave links (Chapter 3.3).

Messages are encrypted either off-line or on-line. With off-line operation, the teleprinter terminal TP (Fig. 136) is supplied with the encrypted text in the form of a perforated tape, for instance. This tape may have been prepared at any other location.

In the case of on-line operation (Fig. 137), the encrypting device EE is frequently interposed between teleprinter terminal TP and data-protection equipment DP. The message is picked up by the data-protection equipment from the teleprinter terminal via the encrypting device.

When the crypto phase is lost in off-line operation, it may be recovered by a manual correction of the position of the tape at the receiving end. On-line encryption, however, is an automatic process so that manual intervention is normally not possible. Protection methods must therefore be employed

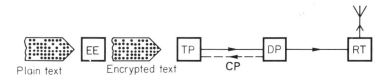

Fig. 136   Off-line encryption (transmitting station only is shown)

EE   Encrypting equipment
TP   Teleprinter terminal, e.g. tape transmitter for call-down operation
DP   Data protection equipment (ARQ or FEC)
RT   Radio transmitter
CP   Calldown pulse

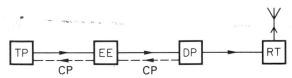

Fig. 137   On-line encryption (transmitting station only is shown)

(ARQ or FEC, see Chapter 7). Establishment of the crypto phase is a different process for on-line and off-line operation. At any rate, an arrangement must be made to the effect that the crypto phase between transmitting and receiving station is ensured.

In off-line operation this arrangement may provide for a starting mark in a tape code (start position both at the transmitting and the receiving end).

In on-line operation this arrangement takes on the form of a preamble inserted ahead of the encrypted text. In this case in particular data protection is a necessity.

10.3 Signal Delay in ARQ Systems

When ARQ terminals are used on shortwave radio links, the round-trip delay must not exceed a certain value. This value must be large enough to permit a repetition process to take place in the overall system. The larger the number of the characters an ARQ terminal is able to store, the longer may be the signal delay. The generally adopted repetition cycles cover either four or eight characters with three or seven characters being stored (see Section 7.5). The permissible total delay, being a function of the repetition cycle employed, has been compiled in Table 14 for the most frequently used telegraph speeds.

In view of the character-by-character call-down of the message the delay time at the subscriber side enters the calculation if the data terminal equipment is not placed immediately adjacent to the ARQ terminal. Table 15 gives the permissible delay times on the subscriber side for a single-channel and for a multichannel data protection equipment.

Table 14 Permissible delay times on transmission path (round-trip delay)

Baud rate of terminal equipment	Repetition cycle in ms			
	Single-channel terminal		Multichannel terminal	
	4-char.	8-char.	4-char.	8-char.
50	135	709	291	875
75	90	472	194	583
100	67	353	146	437
200	33	177	—	—

The delay time of the signals on the transmission path is determined by the length of the radio link and the distance between the transmitting ARQ terminal and the radio transmitter, as well as the distance between radio receiver and the receiving ARQ terminal. The modulators and demodulators also add to the delay time, the major contributors being their filters. Table 16 shows the delay time of the most frequently used systems.

Table 15 Permissible delay times on subscriber side

Baud rate of terminal equipment	Delay times on subscriber side in ms	
	Single-channel terminal	Multichannel terminal
50	135	293
75	90	192
100	62	145
200	31	73

Table 16a Delay times of generally used transmission systems

Class of emission	Baud rate and frequency shift in Hz of modulation and demodulation facilities	Delay time in ms
F1	50 bauds, ± 42.5 Hz	approx. 20
F1	200 bauds, ± 85 Hz	approx. 12
A3J with WTK FM 170	100 bauds, ± 42.5 Hz	approx. 25
A3J with WTK FM 340	200 bauds, ± 85 Hz	approx. 16

Table 16b Delay times of generally used connections

Connection	Delay time in ms
VFT channel FM 120	approx. 30 (± 30 Hz)
VFT channel FM 240	approx. 20 (± 60 Hz)
Carrier circuit (one section)	approx. 8
Shortwave circuit 1000 km	approx. 3
Satellite link (Frankfurt/M−New York)	approx. 280

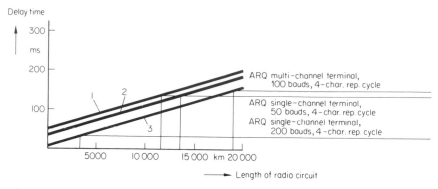

Fig. 138 Maximum length of radio link as a function of its delay

Curve 1   A3J class of emission with WTK FM 170 ($\pm$42.5 Hz)
Curve 2   F1 class of emission, telegraph speed 50 bauds ($\pm$100 Hz)
Curve 3   F1 class of emission, telegraph speed 200 bauds ($\pm$200 Hz)

Figure 138 shows the permissible delay times of ARQ terminals, taking into account the different modulation and demodulation facilities. These figures provide the basis for calculating the maximum allowable distance between a transmitting and a receiving station.

10.4 Error Rate on Shortwave Radio Links

The character error rate to be expected in shortwave radio traffic is shown in Table 17 for several transmission methods. As can be seen, a character error rate of 1×10^{-2} may be anticipated for F1 operation (frequency shift keying with bandwidth matched to modulation rate and frequency swing) over an extended period, if no data protection equipment is provided. Employing space or frequency diversity, an improvement by a factor of about 10 can be achieved. The use of FEC data protection equipment on shortwave radio links reduces the error rate to 2.5×10^{-5}. With ARQ methods the effect of disturbances can be further mitigated so that the error rate ranges between 1×10^{-5} and 1×10^{-6}. This result can still be improved by the use of ARQ and diversity methods.

One possibility of interconnecting transmission terminals, data protection equipment and teleprinters is shown in Fig. 139.

184

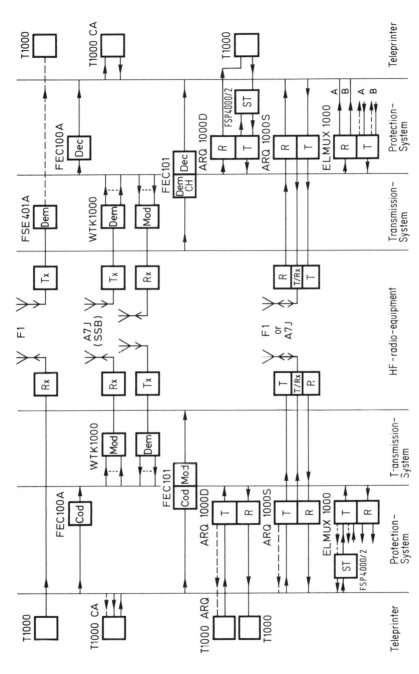

Fig. 139 Possibility of interconnecting transmission terminals, data protection equipment and teleprinters (mode of operation A7J and F1)

Table 17 Typical values of the character error rate to be anticipated for various transmission methods

Transmission method	Character error rate
F 1 operation without data protection	1×10^{-2}
F 1 operation with space or frequency diversity	1×10^{-3}
F 1 operation with data protection (FEC, error correction without backward channel)	2.5×10^{-5}
F 1 operation with data protection (ARQ, error correction by signal repetition)	1×10^{-5} to 1×10^{-6}
F 1 operation with data protection (ARQ and diversity)	1×10^{-6}
A 3J-operation with data protection FEC and channel selection (three WTK channels)	3×10^{-6}

11 Relevant CCIR and CCITT Recommendations

XIV. Plenary Assembly, Kyoto, 1978

1. CCIR

[1] Rec. 106-1   Voice-frequency telegraphy on radio circuits

[2] Rec. 246-2   Frequency-shift keying

[3] Rec. 328-2   Spectra and bandwidths of emissions

[4] Rec. 342-2   Automatic error-correcting system for telegraph signals transmitted over radio circuits

[5] Rec. 343-1   Facsimile transmission of meteorological charts over radio circuits

[6] Rec. 344-2   Standardization of phototelegraph systems for use on combined radio and metallic circuits

[7] Rec. 345   Telegraph distortion

[8] Rec. 346-1   Four-frequency diplex systems

[9] Rec. 436-1   Arrangement of voice-frequency telegraph channels working at a modulation rate of about 100 bauds over HF radio circuits

[10] Rec. 349-2   Frequency stability required for single-sideband, independent-sideband and telegraph systems to make the use of automatic frequency control superfluous

[11] Rec. 456   Data transmission at 1200/600 bit/s over HF circuits when using multichannel voice-frequency telegraph systems and frequency-shift keying

[12] Rep. 195   Prediction of the performance of telegraph systems in terms of bandwidth and signal-to-noise ratios in complete systems

2. CCITT

[13] Rec. R39   Voice-frequency telegraphy on radio circuits

[14] Rec. U20   Telex and Gentex signalling on radio channels (synchronous 7-unit systems affording error correction by automatic repetition)

[15] Rec. U23   Use of radiotelegraph circuits with ARQ-equipment for fully automatic Telex calls charged on the basis of elapsed time

[16] Rec. S12   Conditions which must be satisfied by synchronous systems operating in connection with start-stop teleprinter systems

[17] Rec. S13   Use on radio circuits of 7-unit synchronous systems giving error correction by automatic repetition

[18] Rec. V28   Electrical characteristics for unbalanced double-current interchange circuits

187

12 List of Abbreviations

Abbreviations of Standards Organizations

AEF	Ausschuss für Einheiten und Formelgrössen (Committee for Units and Parameters)
ASA	American Standards Association
BS	British Standard
CCIR	Comité Consultatif International des Radio-Communications (International Consulting Committee on Radio-Communications)
CCITT	Comité Consultatif International Télégraphique et Téléphonique (International Telegraphic and Telephonic Consulting Committee)
CEPT	Conférence Européenne des Administrations des Postes et des Télécommunications (European Conference on the Administration of Postal and Telecommunication Services)
DNA	Deutscher Normenausschuss (German Standards Committee; DIN is its standards symbol)
FNE	Fachnormenausschuss Elektrotechnik (Electrotechnical Committee)
IA	International Alphabet
IEC	International Electrotechnical Commission
ISO	International Organization for Standardization
ITA	International Telegraph Alphabet
NTG	Nachrichtentechnische Gesellschaft im VDE (Communications Organization within the VDE)
SI	International System of Units
UIT (ITU)	Union Internationale des Télécommunications (International Union of Telecommunications)
USA SC II	USA Standard Code for Information Interchange
VDE	Verband Deutscher Elektrotechniker e.V. (Association of German Electrical Engineers)
WMO	World Meteorological Organization

188

General Abbreviations

ARQ	Automatic Request
ELMUX®	ARQ multiplex terminal with automatic error correction
FEC	Forward Error Correction
Gentex	General telegraph exchange
Telex	Teleprinter exchange
UT	Superimposed telegraphy
WTK	Voice frequency telegraph terminal for operation over shortwave radio circuits
MUF	Maximum Usable Frequency
LUF	Lowest Usable Frequency
VO Funk	Vollzugsordnung für den Funkdienst (Radio service regulations)
FOT	Fréquence optimale de travail (Optimum work frequency)
FSK	Frequency shift keying
BU	A ... Letters shift
ZI	1 ... Figures shift
CR	< Carriage return
LF	≡ Line feed
ZWR	Space
R_{12}	Sunspot number (over a period of 12 months)
IFRB	International Frequency Registration Board

13 Selected Bibliography

Bastong, K.: Sonnenflecken und Funkverkehr, Funkschau, Mai 1970, Heft 5, Seiten 137–140.

Beckmann, B.; Menzel, W.; Vilbig, F.: Grenzwellen und Streustrahlung in der Funkausbreitung, Telegraphen-Fernsprech-Funk- und Fernseh-Technik, Band 30, Februar 1941, Heft 2, Seiten 43–52.

Beckmann, B.; Menzel, W.; Vilbig, F.: Über die praktische Bedeutung der Ionosphärenforschung für den Funkdienst, Telegraphen-Fernsprech-Funk- und Fernseh-Technik, Band 29, April 1940, Heft 4, Seiten 106–117.

Beger, H.: Fehlerhäufigkeit von A1- und F1-Telegraphübertragungssystemen insbesondere bei weißem Rauschen, Telefunken-Zeitung, Jahrgang 29, Dezember 1956, Heft 114, Seiten 245–255.

Behrendt, G.: Fernschreibübertragungsversuche in der Handelsschiffahrt, HANSA-Schiffahrt-Schiffbau-Hafen, 105. Jahrgang, 1968, Nr. 19, Seiten 1681–1684.

Besslich, Ph.: Fehlerwahrscheinlichkeit binärer Übertragungen bei Mehrfachempfang und bei frequenzselektivem Schwund, Archiv der Elektrischen Übertragung, Band 17, Juni 1963, Heft 6, Seiten 271–277.

Besslich, Ph.: Fehlerwahrscheinlichkeit binärer Übertragungsverfahren bei Störung durch Rauschen und Schwund, Archiv für Elektronik und Übertragungstechnik, Band 17, April 1963, Heft 4, Seiten 185–197.

Bocker, P.: Datenübertragung – Nachrichtentechnik in Datenverarbeitungs-systemen, Band I und II, Springer-Verlag, Berlin–Heidelberg–New York 1976.

Bode/Laub/Pumpe: Nachrichtenübertragung mit Kurzwellen, Siemens-Zeitschrift 48, Beiheft Nachrichten-Übertragungstechnik, 1974.

Brennau, D. G.: Linear Diversity Combining Techniques, Proceedings of the IRE, June 1959, pages 1075–1102.

Bretschneider, F.: Die Entwicklung der Funktelegraphie im letzten Jahrzehnt, Elektrotechnik 3, 1949, Seiten 145–147.

Buding, H.; Markwitz, W.: Wirksamkeit gespreizter, rekurrenter Codes bei Störungen eines realen Kanals und eines entsprechenden Kanalmodells, Archiv für Elektronik und Übertragungstechnik, Band 34, 1980, Seiten 238–242.

Burger, E.; Hiergeist, M.: Telegrafie-Demodulator FSE 401 für feste und mobile Funkempfangsstationen, Siemens-Zeitschrift 43, 1969, Seiten 783–789.

Burger, E.; Hiergeist, M.: WTK 1000, ein System für Fernschreib- und Datenüber-tragung über Kurzwellen-Funkverbindungen, Siemens-Zeitschrift 48, 1974, Seiten 131–135.

Cohn, D. L.; Levesque, A. H.; Meyn, J. H.; Pierce, A. W.: Performance of selected lock and convolutional codes on a fading HF channel, IEEE Transactions on Information Theory, Vol. IT-14, No. 5, September 1968, pages 627–640.

Das Cupta, J.; Scholz, B.: Vorwärts-Fehlerkorrektursystem mit Kanalauswahl FEC 101 für Fernschreib- und Datenübertragung, Siemens-Zeitschrift 51, 1977, Seiten 613–617.

Dobermann, H.: MUX-Telexplätze für den Fernschreibverkehr mit Übersee, Siemens-Zeitschrift 31, 1957, Seiten 331–335.

Draeger, R. J.: Öllinger, M.: Einkanalgerät ARQ 1a für gesicherte Fernschreib- und Datenübertragung, Siemens-Zeitschrift 41, 1967, Seiten 426–431.

Duuren, H. C. A. van: Error Probability and Transmission Speed on Circuits Using Error Detection and Automatic Repetition of Signals, IRE Transactions on Communications Systems, March 1961, pages 38–50.

Duuren, H. C. A. van: Error control on high frequency radio channels, HET PTT-Bedrijf, Part XIV No. 2, January 1966.

Emberger, E.; Schenk, E.: Meßtechnik für digitale Übertragung in Fernsprech- und Datennetzen, Siemens-Zeitschrift, Jahrgang 43/1969, Beiheft Datenfernverarbeitung, Seiten 83–86.

Emberger, E.: Meßgeräte für Fernschreib- und Datennetze, Siemens-Zeitschrift 48, 1974, Seiten 198–201.

Filter, J. H. J.; Arazi, B.; Thomson, R. G. W.: The Fadeogramm, A Sonogram-Like Display of the Time-Varying Frequency Response of HF-SSB Radio Channels, IEEE Transactions on Communications, Vol. COM-26, No. 6, June 1978, pages 913–917.

Fischer, H.: Faksimile-Wetterkartenübertragung, telcom report 3 (1980) Heft 4, Seiten 352–354.

Fischer, K.; Vesper, W.; Vogt, G.: Übersee-Funkempfangsanlagen für Telegraphie und Einseitenband-Telephonie-Entwicklungsprinzipien und Betriebserfahrungen, Telefunken-Zeitung, Jahrgang 27, März 1954, Heft 103, Seiten 14–26.

Fischer, K.: Übersicht über bekannte und mögliche Funkfernschreibsysteme, Fernmelde-Praxis, Band 37, Februar 1960, Nr. 4, Seiten 121–142.

Franco, A. G.; Wall, M. E.: Coding for error control: an examination of techniques, Electronics, December 1965, pages 70–79.

Frommer, E.; Vogt, K.: Untersuchungen über ein Funkfernschreibsystem mit fehlerberichtigendem Code auf Kurzwellen-Übertragungsstrecken, Fernmelde-Praxis, Band 41/1964, Nr. 23, Seiten 3–11.

Fuchs, E. A.: Telegraphie-Empfangsgerät FSE 30 für Kurzwellen-Funkverbindungen, Siemens-Zeitschrift 35, 1961, Seiten 641–647.

Fuchs, E. A.: "Funk-WT" ein neues Telegrafie-Übertragungssystem in Transistor-Ausführung mit Kanälen verschiedenen Bandbreite, Nachrichtentechnische Zeitschrift 14, 1961, Seiten 419–423.

Fuchs, E.; Pilz, G.: Kurzwellen-Empfangsanlagen für Fernschreib- und Faksimileübertragung, Siemens-Zeitschrift 38, 1964, Seiten 291–293.

Füllung, H.: Fernschreibübertragungstechnik, Verlag: Oldenbourg, München 1957.

Goldberg, B.: 300 kHz–30 MHz MF/HF, IEEE Transactions on Communications Technology, Vol. COM–14, No. 6, December 1966.

Großkopf, J.: Über einige Beobachtungen bei Feldstärkeregistrierungen im Kurzwellenbereich, Telegraphen-Fernsprech-Funk- und Fernseh-Technik, 29. Jahrgang, Mai 1940, Heft 5, Seiten 127–137.

Großkopf, J.; Scholz, M.; Vogt, K.: Korrelationsmessungen im Kurzwellenbereich, Nachrichtentechnische Zeitschrift, Februar 1958, Heft 2, Seiten 91–95.

Großkopf, J.; Heinzelmann, G.; Vogt, K.: Korrelationsmessungen zur Frequenz-Diversity im Kurzwellenbereich, Nachrichtentechnische Zeitschrift, März 1961, Heft 3, Seiten 124–128.

Haase, W.; Neumann, H. J.: Telegraphieempfang im Langwellenbereich, Siemens-Zeitschrift 34, 1960, Seiten 829–835.

Haase, W.; Neumann, H. J.: Ein Duoplex-Schrittordner für Funkfern-schreibsendungen, Nachrichtentechnische Zeitschrift 15, 1962, Seiten 288–292.

Hacks, J.; Grabe, K.: Funktelegrafieempfang in Anlagen mittlerer Größe, Rhode & Schwarz-Mitteilungen, November 1958, Heft 11, Seiten 206–214.

Hagelbarger, D. W.: Recurrent Codes: Easily mechanized Burst-correcting, Binary Codes, The Bell System Technical Journal, July 1959, pages 969–984.

Heidester, R.; Henze, E.: Empfangsverbesserung durch Diversity-Betrieb, Archiv der Elektrischen Übertragung, Band 10, März 1956, Heft 3, Seiten 107–116.

Heidester, R.; Vogt, K.: Untersuchungen zum Diversity-Empfang nach dem Antennen-Auswahl-System, Nachrichtentechnische Zeitschrift, November 1958, Heft 11, Seiten 315–319.

Heinzelmann, G.; Lenhart, B.; Vogt, K.: Gütemeß- und Registriereinrichtung für Funktelegrafielinien, Nachrichtentechnische Zeitschrift, Februar 1969, Heft 2, Seiten 101–106.

Henning, F.: Funkfernschreiben mit selbsttätiger Fehlerkorrektur, Nachrichtentechnische Zeitschrift 9, 1956, Seiten 341–348.

Hiergeist, M.: Telegrafie-Demodulator FSE 401A für ortsfest und mobile Funkstationen, telcom report, 2. Jahrgang, Dezember 1979, Heft 6, Seiten 406–412.

Jendra, H.; Mair, E.: Telex-Auslandsplatz 6b für Draht- und Funkwege, Siemens-Zeitschrift, Jahrgang 40, November 1966, Heft 11, Seiten 782–786.

Kettel, E.: Die Fehlerwahrscheinlichkeit bei binärer Frequenzumtastung, Archiv der Elektrischen Übertragung, Band 22, Juni 1968, Heft 6, Seiten 265–275.

Kirschner, U.: Schnelle und gesicherte Datenfernübertragung, Fernmeldepraxis 44, 1967, Seiten 755–778.

Kohlenberg, A.; Forney, Jr., G. D.: Convolutional Coding for Channels with Memory, IEEE Transactions on Information Theory, Vol. IT-14, No. 5, September 1968.

Kronjäger, W.; Lenhart, B.; Vogt, K.: Über das Raum-Diversity-Empfangsverfahren nach dem Antennen-Auswahl-System, Nachrichtentechnische Zeitschrift, September 1956, Heft 9, Seiten 424–430.

Kronjäger, W.; Lenhart, B.; Vogt, K.: Die Gleichlaufkorrektur von Start-Stop-Fernschreibsystemen bei Funkempfang, Nachrichtentechnische Zeitschrift, April 1957, Heft 4, Seiten 167–174.

Küpfmüller, K.: Telephonie und Mehrfachtelegraphie auf kurzen Wellen, Telefunken-Zeitung 1929, Nr. 53, Seite 23.

Lehnert, J.: Einführung in die Fernschreibtechnik, München, Siemens-Fachbuch 1974.

Leypold, D.; Schucht, P.; Wich, H.: Neue Kurzwellen-Großstationsempfänger für Telegraphie und Telephonie, Frequenz, Band 15, Jahrgang 1961, Heft 2, Seiten 2–9.

Loriol, de, F.: Neue Methoden der drahtlosen Telegraphie, Sende- und Empfangstechnik in der Schweiz, Bullentin SEV 52, (1961) 14, 15. Juli, Seiten 517–523.

Maier, H. P.: Ein stocharstisches Fehlermodell für binäre Datenübertragung auf Kurzwelle, Archiv für Elektronik und Übertragungstechnik, Band 29, Oktober 1975, Heft 10, Seiten 409–415.

Markwitz, W.: Automatische Fehlerkorrektur auf mäßig gestörten Verbindungen in Daten- und Funknetzen, Nachrichtentechnische Zeitschrift 31, 1978, Seiten 274–280.

Markwitz, W.; Niethammer, D.: Vorwärts-Fehlerkorrektursystem FEC 100 für Fernschreib- und Datenübertragung, Siemens-Zeitschrift 48, 1974, Seiten 136–139.

Massey, James, L: Threshold Decoding, Cambridge, Massachusetts: M.I.T. Press, 1963.

McCullough, R. H.: The Binary Regenerative Channel; The Bell System Technical Journal, October 1968, pages 1713–1735.

McManamon, P.; Janc, R.: An Experimental Comparison of Nondiversity and Dual Frequency Diversity for HF FSK Modulation, IEEE Transactions on Communication Technology, December 1968, No. 16–6, pages 837–839.

Meinke, H.: Gundlach, F. W.: Taschenbuch der Hochfrequenztechnik, Berlin–Göttingen–Heidelberg, 2. Aufl., 1962, Springer-Verlag.

Meissner, H.: Telegrafie-Einrichtung im neuen Kurzwellenempfänger für Groß-Funkstationen, Frequenz 15, 1961, Seiten 56–58.

Menzel, W.: Ionosphärische Einflüsse auf die Wellenausbreitung (Grundlagen des Funkwetterdienstes), Der Fernmelde-Ingenieur, 7. Jahrgang, November 1953, Heft 11, Seiten 1–32.

Mücke, W.: Die Bedeutung der Antenne für Kurzwellenverbindungen, Fernmelde-Praxis, Band 38, Juni 1961, Heft 12, Seiten 441–464.

Müller, F.: Das Antennenauswahlgerät H305, Fernmelde-Praxis, Band 42/1965, Nr. 13, Seiten 512–518.

Nestel, W.: Wellenfragen als internationales und technisches Problem, Telefunken-Zeitung, Jahrgang 30, September 1957, Heft 117, Seiten 161–173.

Neumann, H. J.: Wechselstromtelegraphiesystem für Kurzwellen-Telefonieverbindungen FM-WTK 3/6, Nachrichtentechnische Zeitschrift 11, 1958, Seiten 510–514.

Paetsch, W.; Vogt, W.: Elmux 1000, ein neues ARQ-Multiplexsystem für Funkfernschreiben, Siemens-Zeitschrift 45, 1971, Seiten 123–129.

Peterson, W. W.: Prüfbare und korrigierbare Codes, Verlag R. Oldenbourg, München und Wien 1967. (Deutsche Ausgabe).

Peterson, W. W.; Weldon, E. J.: Error-correcting codes, Cambridge, Massachusetts: M.I.T. Press 1972 (English edition).

Pierce, J. N.: Theoretical Diversity Improvement in Frequency-Shift-Keying, Proceedings of the IRE, 1958, pages 903–910.

Reche, K.; Arzmaier, A.; Zimmermann, R.: Verringerung der Fehleranfälligkeit drahtloser Telegraphierwege durch Maßnahmen im Niederfrequenzteil der Übertragungssysteme, Telefunken-Zeitung, 1939, Nr. 80, Seite 53.

Retting, H.; Vogt, K.: Schwunddauer und Schwundhäufigkeit bei Kurzwellenübertragungsstrecken, Nachrichtentechnische Zeitschrift, Jahrgang 17, Februar 1964, Heft 2, Seiten 57–62.

Retting, H.: Untersuchungen über den Zusammenhang zwischen der Zahl der Wiederholungen und der Fehlerhäufigkeit bei Funkfernschreibsystemen mit automatischer Rückfrageeinrichtung, Nachrichtentechnische Zeitschrift, 19. Jahrgang, Januar 1966, Heft 1, Seiten 7–10.

Retting, H.: Untersuchungen über die Meßmöglichkeiten und die Größe des erforderlichen Meßintervalls zur Bestimmung des Wirksamkeitsfaktors von Funktelegrafiesystemen, Nachrichtentechnische Zeitschrift, 19. Jahrgang, Januar 1966, Heft 1, Seiten 11–14.

Retting, H.: Untersuchungen an Funk-Wechselstromtelegrafiesystemen, Nachrichtentechnische Zeitschrift, 19. Jahrgang, September 1966, Heft 9, Seiten 513–526.

Roßberg, E.: Stand und Entwicklungstendenzen des interkontinentalen Fernschreibverkehrs, Nachrichtentechnische Zeitschrift, 19. Jahrgang, Dezember 1966, Heft 12, Seiten 726–728.

Rudolph, H.; Bochmann, K.: Ein elektronisches Multiplex-Fernschreibsystem mit automatischer Fehlerkorrektur – ELMUX, 2/4 D7, Siemens-Zeitschrift 33, 1959, Seite 534–541.

Schönhammer, K.; Voss, H. H.: Fernschreibübertragungstechnik, München–Wien: Oldenbourg 1966.

Spiegel, H.: Ein Zeitmultiplex-Fernschreibsystem mit Fehlerkorrektur für Funkverbindungen, Siemens-Zeitschrift 29, 1955, Seiten 364–368.

Steinbuch, K.: Taschenbuch der Nachrichtenverarbeitung, Springer-Verlag, Berlin–Heidelberg–New York, 1962.

Süßmann, P.: Die weltweite Ausbreitung von Kurzwellen über die Ionosphäre und ihre Vorhersage, Nachrichtentechnische Zeitschrift, Jahrgang 29, Mai 1976, Heft. 5, Seiten 394–399.

Swoboda, J.: Ein statistisches Modell für die Fehler bei binärer Datenübertragung auf Fernsprechkanälen. Archiv für Elektronik und Übertragungstechnik. Jahrgang 23, 1969, Seiten 313–322.

Thierbach, D.: Gleichzeitige Telegraphie und Telephonie auf Kurzwellenverbindungen (u.a. Multiplexsystem), Telefunken-Zeitung 1932, Nr. 60/61, Seiten 36/19.

Thierbach, D.; Sedlmayer, J.: Ein neuzeitliches Telegraphierverfahren im Kurzwellenverkehr, VDE-Fachberichte 6, 1934, Seite 118.

Tsai, S.: Evaluation of Burst Error Correcting Codes Using a Simple Partitioned Markov Chain Model, IEEE Transactions on Communications Technology, 1973, pages 1031–1034.

Voss, H. H.: Realisierbare Tiefpässe und Bandpässe minimaler Phase mit geebneter Laufzeit und – aperiodischen Einschwingverfahren, Frequenz 8, 1954, Seiten 97–102.

Voss, H. H.: Eigenschaften von Telegraphie-Übertragungssystemen mit Frequenzmodulation bei Störungen durch Rauschen, Frequenz 12, 1958, Seiten 31–37.

Voss, H. H.; Neumann, H. J.: Telegraphieübertragung im Kurzwellenbereich, Nachrichtentechnische Zeitschrift 12, 1959, Seiten 343–347.

Voss, H. H.: Fernschreibübertragung auf Funkverbindungen, Siemens-Zeitschrift 34, 1960, Seiten 463–469.

Wellhausen, H.-W.; Martin, D.: Fehlerhäufigkeitsmessungen, Nachrichtentechnische Zeitschrift, 24. Jahrgang, November 1971, Heft 11, Seiten 553–557.

Wiesner, L.: Funkfernschreiben auf Kurzwellen mit Telegrafie-Empfangstastgerät FSE 30, Fernmeldepraxis 42, 1965, Seiten 287–301.

Wiesner, L.: Funkfernschreibanlage für gesicherte Nachrichtenübertragung auf Kurzwellenverbindungen, Fernmeldepraxis 47, 1970, Seiten 633–644.

Wiesner, L.: Gesicherte Fernschreib- und Datenübertragung auf Kurzwellenfunkverbindungen mit dem neuen ARQ-Simplex-System ARQ 1000 S, telcom report, 1980, Seiten 441–445.

Wild, A.: MKS 4000, ein Magnetkernspeicher für Fernschreib- und Datennetze, Siemens-Zeitschrift 41, 1967, Seiten 566–571.

Wüsteney, H.; Henning, F.: Typendrucktelegraph für den drahtlosen Betrieb, Telefunken-Zeitung 1934, Nr. 69, Seite 15.

Wu, W. W.: New Convolutional Codes, Part I IEEE Transactions on Communications Technology, COM-23, 1975, pages 942–955.

Index